MINISTRY OF
AGRICULTURE, FISHERIES AND FOOD

Lime and Liming

Bulletin 35

LONDON
HER MAJESTY'S STATIONERY OFFICE
1969

First published 1931
Sixth edition 1969

The Ministry does not accept responsibility for any of the private or trade advertisements included in this publication

SBN 11 240335 2

Contents

Foreword

LIMING is fundamental in maintaining soil fertility, and any serious shortage in the soil needs to be put right at the outset. A century ago lime was often given without adequate manuring, and though in naturally fertile soils the response was good, such liming too often accentuated nutritional deficiencies. Modern practice emphasizes the importance of liming in conjunction with proper fertilizer treatment.

It is the purpose of this bulletin to promote better understanding of the value of liming and, by providing information based on scientific knowledge and practical experience, to help the farmer in working out his liming policy. This sixth edition is based largely on the fifth edition but the information has been brought up to date to take account of the changes in practice. The new basis of evaluation of soil liming materials in terms of neutralizing value is explained and particulars from the new Regulations under the Fertilisers and Feeding Stuffs Act are given.

This edition has been prepared by the Soil Science Publications Committee of the National Agricultural Advisory Service, under the chairmanship of Mr. J. B. E. Patterson, M.Sc., F.R.I.C., with special help from Mr. G. L. Gray, B.Agr. The Committee also wishes to acknowledge the help given by technical officers of the Ministry in the specialist sections of the bulletin. The general editor is Mr. Ronald Ede.

Ministry of Agriculture, Fisheries and Food

June 1968

Lime in Agriculture

LIMING has been a major factor in the pattern of British agriculture since time immemorial and there are many historical references to its use on arable soils. The very large dressings—chiefly of chalk—which were applied in the nineteenth century created reserves which lasted for many years. As crop yields increased with the introduction of better varieties and higher standards of farming, this capital was gradually used up, not only under arable farming but also where grassland and livestock management were improved and dairying became more intensive.

Since the late 1930s, the Government has encouraged the greater use of lime to restore fertility to neglected land and assist in greater production.

THE NEED FOR LIME

The importance of lime in soil fertility is widely recognized and throughout large areas the deficiency has been made good and substantial reserves have been built up. There are still many farms on which soil acidity is a problem. On farms where the lime status is adequate, not only would it be uneconomic to apply lime but it would increase the risk of troubles from over-liming. Where, however, it is still necessary, as on light sandy soils, farmers need not be deterred by the fear that heavy dressings must be applied at great cost: a ton or two per acre is often enough.

Great progress has been made in the last few years in the study of the effects of lime on soils and crops, and analytical methods have been devised for finding out whether it is needed and, if so, how much should be applied. There is little excuse nowadays for either over-liming or under-liming.

LIMING POLICY AND PRACTICE

During the 1930s, the Government recognized that in the event of war Britain would become dependent largely on home-grown foodstuffs, and the building up of soil fertility would become a front-line need. It therefore started to subsidize the use of lime through the medium of the Land Fertility Committee, which was created under the authority of the Agriculture Act, 1937. The subsidy scheme, which came into operation in 1937, provided at the outset for a contribution of up to half the cost of the lime and half the cost of transporting it to the place where it would be used. Latterly, the principle of a percentage contribution of cost has been abandoned in favour of a flat rate of contribution by grades of lime including the cost of spreading. In addition, contributions are made towards the cost of transport.

The main objective of this policy is to improve the fertility of the land by raising the lime status of the soil to an optimum level, based usually on a pH value of about 6·5 (fen peats 5·8). To this end, not only has the subsidy scheme continued in force, but increased production of lime has met the increased demand for lime.

PRACTICAL EFFECTS

Before the introduction of this scheme in 1937, farmers in England and Wales were using about 400,000 tons of lime annually, mainly in the form of burnt lime and its derivatives. At present, farmers are taking about five

(109339) A*

million tons of liming materials annually, mostly in the form of ground carbonate and screened chalks.

Technical developments in production, mainly directed towards the manufacture of inexpensive materials, have been successful in both chalk and limestone districts. The effect of the tendency for farmers to use ground limestone in preference to burnt lime has been the creation of a large demand for the former and the development within the industry of a productive capacity to meet this demand throughout the year. The Government gave financial assistance during the early years of the war towards the erection of limestone grinding plant. Since then, the industry has taken steps to meet increasing demands by providing additional productive capacity wherever natural resources have permitted.

As the demands on the more conventional types of liming materials have increased, users have turned to other sources of readily available forms of calcium carbonate, in particular the by-products of the chemical industries. The products are mainly in the form of precipitated carbonates, which at peak demand sell readily, and within limits of quality and delivery costs have proved themselves useful and cheap sources of supply.

The greater use of liming materials has been accompanied by the development of mechanical spreading by the supplier or other contractor. At present it is estimated that nearly 80 per cent of the materials coming from the works are spread in this way. The introduction of spreading machines has resulted in the production of new types of liming materials: for example, some manufacturers now process burnt lime by roughly grinding it so that it can be spread mechanically.

The introduction of spreading services has also resulted in a reduction in demand for liming materials packed in paper and hessian bags. Formerly, many farmers were taking delivery in bags which increased the cost. At first the cost of paper bags was about £1 per ton of product; by spreading mechanically it has been possible to save bagging costs of bags, and so more than offset the cost of the spreading itself.

The long-term effect of the subsidy scheme has been, particularly in the eastern counties, to bring the lime status of the soil to a more satisfactory and economic level. On many farms, it is no longer necessary to apply large quantities at one time, but liming is now part of the normal rotational treatment of the soil. In other words, the time is drawing nearer when the practice of liming can best be described as one of little and often.

In building up lime status of the soil, farmers should have regard to the cost per acre—a significant factor in all liming. In areas where alternative supplies are available there seems to be an increasing tendency for farmers to discriminate on the basis of cost per acre of the requisite dressing delivered and spread. This practice is to be encouraged both from the point of view of the farmer and the Exchequer.

Loss of Lime from Land

LOSSES of lime from the soil are due to various causes, of which the following are important: removal in crops or livestock sold off the land, losses in drainage water, the effects of acid-charged smoke and the effects of fertilizers. Of all these possible causes, leaching from the soil in drainage water is undoubtedly the most important, being greatest in well-drained uncropped land and outstanding in smoke-polluted areas.

LIME IN THE SOIL

The natural lime supply of the soil obviously depends partly on the chemical composition and partly on the physical character of the parent rocks. Over chalk and some of the softer limestones and over boulder clays containing chalk, the soils generally, but by no means invariably, contain lime. On the other hand, some formations, notably the Millstone Grit, Coal Measures, the Triassic Sandstones, and some of the glacial sands furnish very little lime, and the soils derived from them are generally acid. Even where the rock contains plenty of calcium carbonate the overlying soil may be acid.

In the humid climate of Britain lime is continually being washed out of the soil in drainage, mainly in the form of calcium bicarbonate. If no lime were added, the soil would eventually become almost entirely depleted of available lime. With soils naturally rich in it as, for example, many of the soils on the chalk, this depletion would take centuries.

On the other hand, in many wet upland regions there are soils almost completely devoid of lime. Such soils are said to be acid, and when lime is added to them the so-called soil colloids, or in other words the humus and clay, behave as weak insoluble acids and are, therefore, able to react with a base, such as quicklime (calcium oxide), hydrated lime (calcium hydroxide), carbonate of lime (calcium carbonate) or magnesian limestone (magnesium and calcium carbonates). The lime thus combined is often called 'exchangeable lime' on account of the readiness with which it takes part in chemical reactions. It is in fact the 'active lime' of the soil.

When lime dressings of any kind are added to the soil the capacity of the soil to combine with lime is satisfied first, and any excess remains over in the form of carbonate of lime. So long as the soil contains a reserve of calcium carbonate, well distributed through the soil and not merely as occasional hard fragments, the soil may be regarded as saturated with active or exchangeable lime.

The absence of reserve calcium carbonate does not necessarily mean, however, that the soil must be limed. Many of the most fertile soils are not fully saturated with lime. Although they are slightly acid and have no reserves of carbonate, they have sufficient active lime to meet all requirements, and need only occasional light dressings to make good the inevitable losses.

Water draining through the soil necessarily removes some lime. So long as it has reserves of calcium carbonate these carry the loss, and the soil remains saturated and neutral. When the reserves have gone the loss falls on the exchangeable lime, and the soil then becomes progressively more acid. At first this is of little consequence, but as the lime shortage and the

(109339) A* 2

associated acidity increase characteristic crop failures set in and fertility falls until the soil is recognized as 'sour'.

CROP LOSSES

Since lime is an essential plant nutrient, it enters into the composition of all crops, and the removal of crops, either directly or through the feeding of livestock, leads to depletion of the soil's lime reserves. With cash crops, very little if any lime is returned to the soil. With crops utilized for stock feeding, only a part is returned to the soil, not necessarily in the same field if the crop is not fed on the land. These losses are, however, generally small in comparison with losses from other causes and are only substantial with kale, leguminous hay crops and some leafy fodders. The amount of lime removed from the soil by farm crops is given in Table 1.

TABLE 1

Calcium content of farm crops grown on fertile soils
Pounds of lime (calcium carbonate) in one ton

		lb
Wheat	grain	2
	straw	11
Oats	grain	6
	straw	14
Barley	grain	2½
	straw	20
Peas		4
Pea straw		64
Potatoes	tubers	2
Sugar beet	roots	2½
	tops	13
Mangels	roots	1
Swedes	roots	3½
Kale	(marrow stem)	18
Meadow	hay	39
Seeds	hay	8
Lucerne	hay (2 cuts)	110
Grass meal		48

The following table shows the net loss of lime from four typical farming systems. Allowance has been made in the calculation for the loss of lime in the buildings during the winter when making and storing farmyard manure and the amount of calcium obtained from purchased feedingstuffs.

TABLE 2

Amounts of ground limestone removed from 100 acres of farmland from the sale of produce

Farming system		Ground limestone cwt per annum
Arable	70 acres cereals, 20 acres sugar beet, 10 acres potatoes	
	all straw retained	11·0
	all straw sold	19·2
Mixed	30 acres cereals, all straw retained, 10 acres cash roots, 60 acres temporary grass, milk, calves and culled stock sold from 45 milking cows, plus followers . . .	35·0
Dairy	All grass; milk, calves and culled stock from herd of 65 cows plus followers	45·0
Grass Drying	All grass, very high nitrogen fertilizer usage; all herbage sold	174·0

DRAINAGE LOSSES

Rain and soil water are charged with carbon dioxide (carbonic acid), which dissolves the lime contained in the soil. Some of this dissolved lime is always finding its way down into the drainage system and thus being lost from the soil. The rate of loss depends on the rainfall, the texture of the soil and the amount of calcium in the soil and varies from 1–8 cwt of calcium carbonate per acre annually. The loss is high from well-drained calcareous or neutral soils and is higher under arable cropping than under grass. Under most systems of arable farming the land is bare for some part of the year, especially in winter when the soil tends to be at its wettest; it is then that land is particularly exposed to leaching, which is greatest in uncropped ground.

AIR POLLUTION

The lime loss in drainage is even greater when rain-water contains other acids besides carbonic acid. In industrial districts and near large cities sulphur dioxide arising from the combustion of fuels is produced in large quantities and some of it is washed into the soil by rain. This acid substance greatly increases the loss of lime and in badly polluted areas really good grassland has been known to become practically derelict in ten years. Some of the damage may be done directly by contact of the air pollutants (for example, sulphur dioxide or smoke, or both) but the main damage to farmland is due to the rapid removal of lime from the soil by the sulphur acids and this can be mitigated by timely applications of lime. The impact of the Clean Air Act has greatly reduced the amount of air pollution.

FERTILIZERS AND MANURES

Many farmers believe that the use of artificial fertilizers causes soil acidity, but in fact very few fertilizers influence it, and those that do have much less effect than is frequently supposed. When fertilizers containing sulphates, chlorides or nitrates are added to a soil, these parts of the fertilizer combine with some of the calcium, which may then be washed out of the soil. It is, however, principally salts of ammonia that increase the loss of lime by forming soluble calcium salts. Other fertilizers leave basic residues that slightly reduce the drain on soil lime. The more important effects are dealt with in the following paragraphs.

Nitrogen

Most forms of nitrogenous fertilizer are based on ammonia, such as sulphate of ammonia, ammonium nitrate and ammonium phosphate. Soil bacteria change ammonia into nitrate which combines with the soil calcium to form calcium nitrate which may be lost in drainage water. Urea is also converted into nitrate in the same way. Calcium sulphate, which is soluble in water is also formed from sulphate of ammonia.

Ammonium salts in compound fertilizers also increase the loss, but the ammonium phosphate in some concentrated compound fertilizers has less effect than ammonium sulphate. Ammonium-nitrate lime fertilizers tend to leave basic residues which slightly reduce the drain on the soil lime. These fertilizers add the equivalent of about ½ cwt calcium carbonate for every cwt of fertilizer. It is partly for this reason that these fertilizers are often used on soils deficient in lime.

Organic Manures

Farmyard manure, composts and other forms of crop residues return to the land some of the calcium compounds present in the plants from which they are derived. With the usual rate of dressing on farmland, the gain would be about 1 cwt of calcium carbonate per acre. On the other hand, fresh poultry manure contains insufficient lime to balance losses caused by its ammonia content, and it increases the loss of soil lime.

Phosphates

In spite of frequent statements to the contrary, superphosphate does *not* increase the loss of lime from the soil. Of the other phosphatic fertilizers, basic slag contains calcium compounds which tend to reduce soil acidity. In liming value, 1 cwt of basic slag is equivalent to about ½ cwt of ground limestone; it is, therefore, important not to exaggerate its effectiveness, and still less to assume that a few cwt of basic slag at infrequent intervals can do the work of the tons of lime that may be needed on really acid land.

SOIL CONDITIONS

Two key facts are that soil acidity depends on the lack of saturation with lime, and that soils differ widely in the amounts of exchangeable lime which they hold when saturated. Light, sandy soils have the lowest capacity for holding exchangeable lime, and, as they drain freely, they rapidly lose it. It is on these soils that troubles from sourness are most common and most acute, but fortunately they are the most easily remedied. Since comparatively small amounts of lime will raise the exchangeable lime content towards saturation point, light dressings generally suffice to restore very acid sandy soils to fertility. By contrast, heavy soils or those rich in humus can hold very much larger quantities of exchangeable lime, and shortage sets in much more slowly. Such soils can remain fertile for many years, even though they contain no reserves of carbonate. When, however, the acidity passes the danger point and important crops begin to fail, heavy dressings may be needed to secure good crops.

The rate of loss of lime from soil, therefore, depends on (a) the nature of the soil, and (b) the amount of calcium carbonate and exchangeable lime present in it (see page 3). The relative proportions of sand and clay as well as of organic matter influence the retention of lime, losses being most rapid from sands devoid of organic matter (see Soil Texture on page 10). Other things being equal, the rate of loss is highest from soils containing sufficient lime to be able to hold reserves of carbonate, and falls off as the content of active or exchangeable lime is reduced. It becomes very low on soils which have already lost most of their active lime and are in consequence noticeably acid.

When the cropping of very acid soils is restricted to the more resistant crops, such as potatoes and oats, the small amount of lime remaining is lost very slowly, and only an occasional light liming is required to maintain the lime content at this low level. When, however, it is desired to change over to more sensitive crops, for example, to introduce sugar beet or improve the leys, a heavy initial dressing of lime is needed to make the change possible, and further dressings must follow regularly at a convenient point in the rotation to meet the increased rates of loss at the new level of lime content.

Since lime moves downwards in drainage water, it is quite possible for the surface soil to become acid, even when it overlies a calcareous subsoil. Thus, thin soils over chalk may become so exhausted of lime as to allow Club Root (Finger-and-Toe) to appear in swedes and turnips—and to show other signs of acidity—despite the fact that there is free chalk a few inches below the root range of young plants. Again, the movement of water in the deeper subsoil tends to be uneven, and some patches of land are denuded of lime more rapidly than others; this may lead to patchiness in the crop stand and to the appearance of weeds and diseases as the lime status of the soil approaches the danger point.

Measuring Lime Requirement

SOIL pH

THE degree or intensity of soil acidity and alkalinity is measured on the pH scale; at pH 7 a soil is neutral—neither acid or alkaline. Any figure below 7 denotes that the soil is acid and the lower the figure the more acid the soil is. Figures above 7 indicate that the soil is alkaline; higher figures indicate higher alkalinities. Most of the cultivated soils in this country fall between pH 4·0 and pH 8·0 and by far the greatest number have pH values between 5·5 and 7·5. To those unfamiliar with the use of pH figures, it should be explained that there is no very simple and obvious arithmetical connection between the figures for pH and the amount of lime (if any) required. In fact, between the extremes of 4·0 and 8·0 lies the whole range and complexity of need for lime of the crops and soils of this country, and the interpretation of pH in terms of lime requirement is a matter for the judgment of the expert.

Although plants differ in their tolerance to acidity, most crops thrive well when the pH is about 6·5. This is also the pH value at which most nutrient mineral constituents in the soil are readily available to plants.

Measurement of soil pH can be carried out by the use of appropriate indicator solutions. Indicators are chemical compounds that change colour with small changes in pH. No one indicator gives a complete range of pH values, but it is possible to have a mixture of indicators to cover the necessary range for soil testing. This is useful in the field for giving a rough and ready measure of the extent and distribution of soil acidity, as a preliminary to sampling the soil for the determination of pH and lime requirement in the laboratory. The ranges of pH and colour changes, obtained with one such indicator mixture in frequent use, are approximately as follows:

Alkaline	pH 7·5 and over	blue
Neutral or near neutral	pH 7·5–6·5	bluish green
Slightly acid	pH 6·5–6·0	yellowish green
Moderately acid	pH 6·0–5·5	orange yellow
Acid and very acid	pH 5·5–4·5	reddish orange
Extremely acid	pH 4·5 and under	deep red

Too much reliance should not be placed on field tests with soil pH indicators, for quite apart from serious difficulties with red and other coloured soils and with organic soils, the colour gradation is continuous and not sharp. The results can be very misleading if the tests are not carried out with care and understanding of the pitfalls. Farmers and others using indicators in the field should remember that a mere change in colour of the indicator from blue to bluish green is no indication of a need for lime. It is only when the green becomes faint and the yellow shade appears that liming is needed. Overliming has resulted from attaching too much importance to a slight change in colour.

In the laboratory, indicator solutions have been entirely superseded by the use of direct-reading electrometric meters with soil suspensions in water or salt solutions, the pH being read directly from the scale of the meter.

EFFECT OF SOIL TEXTURE

The pH value of a soil is not itself a measure of the amount of lime needed to correct acidity, but in conjunction with soil texture it is a useful guide to lime requirement. To assess lime need by this means, however, it is essential to classify soil texture, as the lime requirement of soils with the same pH progressively increases from sandy to clay soils. The following example will help to make this clear: if a sandy soil of a given pH value had a lime requirement of about 2 tons calcium carbonate per acre, a silty soil of the same pH might require 3 tons, and a clay perhaps 4 tons.

Although this method is useful in the absence of laboratory measurements, it can only be used satisfactorily by those with sufficient experience to place a soil in its proper textural class. It should also be realized that it is of little use with peats and other highly organic soils.

LIME REQUIREMENTS

Laboratory methods of estimating lime requirement more directly are used, such as the measurement of the amount of calcium absorbed by a soil from a solution of calcium salt in order to bring the pH up to 6·5 (6·0 for permanent grass and 5·8 for fen peats). The results are expressed in cwt of calcium carbonate per acre.

Sometimes the amount of exchangeable calcium in the soil is used as a basis for estimating lime requirement, but this method also depends on assessing the soil texture.

Laboratory tests of lime requirements are carried out under ideal conditions where an equilibrium is set up in a comparatively short time between the soil and calcium in solution. It is impossible to approach these conditions in the field, since lime spreading, however well done, is never really uniform over the whole of the ground. For this reason, a field factor is used to adjust the laboratory results.

Effects of Liming

LIME plays many parts in the soil and is an outstanding factor in soil fertility. The extent and nature of the benefits depend on the need for lime, the character of the soil, the crops being grown, the method and frequency of lime applications and the climate of the area. In the drier parts of the country larger dressings must often be used than in areas of high rainfall. The more important effects can be considered in relation to (i) soil acidity, (ii) soil organisms, (iii) soil structure, (iv) plant nutrients, and (v) crops and livestock.

SOIL ACIDITY

Unless lime is present, soils tend to become progressively more acid, particularly soils naturally rich in organic matter or lacking in clay.

Acid soils contain both 'soluble' and 'insoluble' acids. The soluble acids are mainly carbonic acid and organic acids derived from the decomposition of organic matter, and mineral acids derived from fertilizers added to the soil, or in industrial areas from atmospheric pollution. These acids, being soluble in water, do not normally accumulate in well-drained soil, but combine with lime or are leached out by rainfall. The insoluble acids are not leached out and their presence results in the soil becoming more acid and remaining so.

In acid soils, hydrogen, which is an essential part of every acid, is combined with the surface of the fine particles of clay and well decomposed organic matter—the so-called soil colloids. The fraction of the total surface of the colloids which combines with hydrogen more or less determines the intensity of the acidity. When a liming material is added to an acid soil under suitable conditions the calcium and magnesium it contains reduce acidity by replacing the exchangeable hydrogen on the surface of the soil colloids, and also by combining with any soluble acid present. It is in this way that soil acids become neutralized. At pH 6·5–7·0 the small amount of exchangeable hydrogen left has the useful function of keeping phosphate available without being otherwise harmful to crops.

In addition to calcium and magnesium, other bases such as potassium, sodium and ammonia can replace or be exchanged for the hydrogen existing in the surface of the colloids, and are referred to as exchangeable bases. The total capacity of the soil to take up bases in this manner is referred to as the base-exchange capacity. When the exchange of bases for hydrogen is complete, the base-exchange capacity is said to be saturated. At this point complete neutralization of soil acidity occurs, generally when the pH of the soil is about 7. Since the majority of crops grow best on soils that have a pH value slightly below 7, it is not always desirable to neutralize soil acidity completely.

SOIL ORGANISMS

The soil is inhabited by vast numbers of microscopic organisms—bacteria, fungi, algae and protozoa. The breaking down of plant and animal residues with the formation of humus and the setting free of simple plant nutrients is dependent on their activities. Many soil micro-organisms, however, are not concerned with these essential processes—some indeed are harmful, causing plant diseases or preying upon the beneficial organisms, or impoverishing the soil. The beneficial organisms are all influenced by soil acidity and liming.

The addition of fresh organic matter to the soil causes a great increase in the number of micro-organisms when soil conditions are favourable. In the course of the resulting decomposition, carbonic and other acids are formed. If the amounts of free and exchangeable calcium and other bases are insufficient to neutralize these acids, soil acidity increases until the growth and activity of the micro-organisms is interfered with, even though there may still be some fresh organic matter remaining in the soil. The addition of lime, by reducing acidity, encourages the soil organisms and tends to permit decomposition to proceed to completion, with the accompanying liberation of nutrients. Soil bacteria are generally more adversely affected than fungi by acidity. For this reason, the activity of bacteria tends to predominate under neutral or slightly acid conditions, and that of fungi in the more acid soils.

The development of a mat on neglected grassland under acid conditions is a good illustration of the relationship between soil lime and the activities of soil organisms. In grassland, worms play some part in the removal and digestion of old dead herbage and root fibre; the beneficial bacteria continue the process and, ultimately, release useful plant nutrients. Under acid conditions, the activity of worms as well as of bacteria is reduced: dead herbage accumulates, soil nutrients become locked up in undecayed organic matter, aeration suffers—and as a result a surface mat builds up and the sward rapidly deteriorates. Liming combined with other renovation treatments points the way to improved fertility and better herbage.

The effect of liming on the incidence of most plant diseases is not very great, but a few diseases, especially Club Root (Finger-and-Toe) in turnips and other cruciferous crops and Brown Foot Rot of cereals, occur much more frequently on lime-deficient soils and may, to a large extent, be controlled by liming. On the other hand, diseases such as Take-all of wheat and barley and Common Scab of potatoes occur principally on soils well supplied with lime.

SOIL STRUCTURE

It is widely held by farmers that liming improves the structure of heavy soils, reduces stickiness, lightens cultivations and makes it easier to break down clods to obtain a satisfactory tilth. In the days of horse cultivations, it was often said that after liming 'four-horse land' became 'three-horse land' or even less.

On the whole, experimental evidence fails to confirm these direct benefits of liming on soil structure. It is well known that clay suspensions can be flocculated by calcium salts, and this led to the view that the beneficial physical effects of lime were due to flocculation of the soil colloids. Flocculation may sometimes be necessary as a preliminary to soil crumb formation, but the particles still need binding together before crumb can be formed.

From the point of view of soil structure, two effects of the addition of lime should be distinguished: (i) the effect of the amount of calcium in the base-exchange complex, (ii) the effect of excess calcium carbonate. As a result of the liming of acid soils, calcium replaces, by base exchange, aluminium combined with the clay particles. This in itself is not found to improve the stability of the soil crumbs or to improve tilth, and exceptions are possibly due to secondary effects on the activity of soil organisms. A beneficial effect from the addition of excess calcium carbonate has also not been generally established. The effect of excess carbonate may be to reduce the size of the clods and possibly to make them more stable, which would be advantageous on clay soils. This appears to happen on some clay soils when free lime is present as calcium carbonate, but on the other hand some of the Gault Clay soils have a high natural calcium carbonate content and yet they are among the most difficult soils to work.

Many of the beneficial effects of lime on soil structure may well prove to be indirect. For example, the addition of lime to an acid soil increases the population of bacteria responsible for breaking down organic matter. The decomposition of organic matter is believed to assist soil crumb formation, possibly through the formation of gum-like substances which act as cements, or of soluble humic materials which coat the soil crumbs and bind them on drying.

It is also known that the development of plant roots and the protection of the soil surface by vegetation have beneficial effects on soil structure—well exemplified in grassland, including leys, brought under the plough. Since effective liming promotes a greater development of vegetation, the benefit to soil structure may be partly due to these indirect effects on plant growth and organic matter production.

It must be concluded, however, that the master key to the secret of soil structure and the respective parts played by lime and organic matter in soil improvement has yet to be discovered.

PLANT NUTRIENTS

Liming affects the uptake by crops of both major and minor plant nutrients; it may also affect the toxicity to plants of some of these nutrients, according to the circumstances.

The trend of changes in the availability of plant nutrients under the influence of soil reaction (pH) is illustrated in the diagram (Fig. 1). For any given nutrient element, the wider the band in the diagram the more favourable the soil reaction is for an abundant supply of the nutrient in available form.

With some elements these effects are mainly associated with changes in availability due to conversion to more soluble forms. Some become more available as soil acidity increases, even to the extent of becoming toxic to crops (e.g., manganese) under extreme conditions. Others become particularly

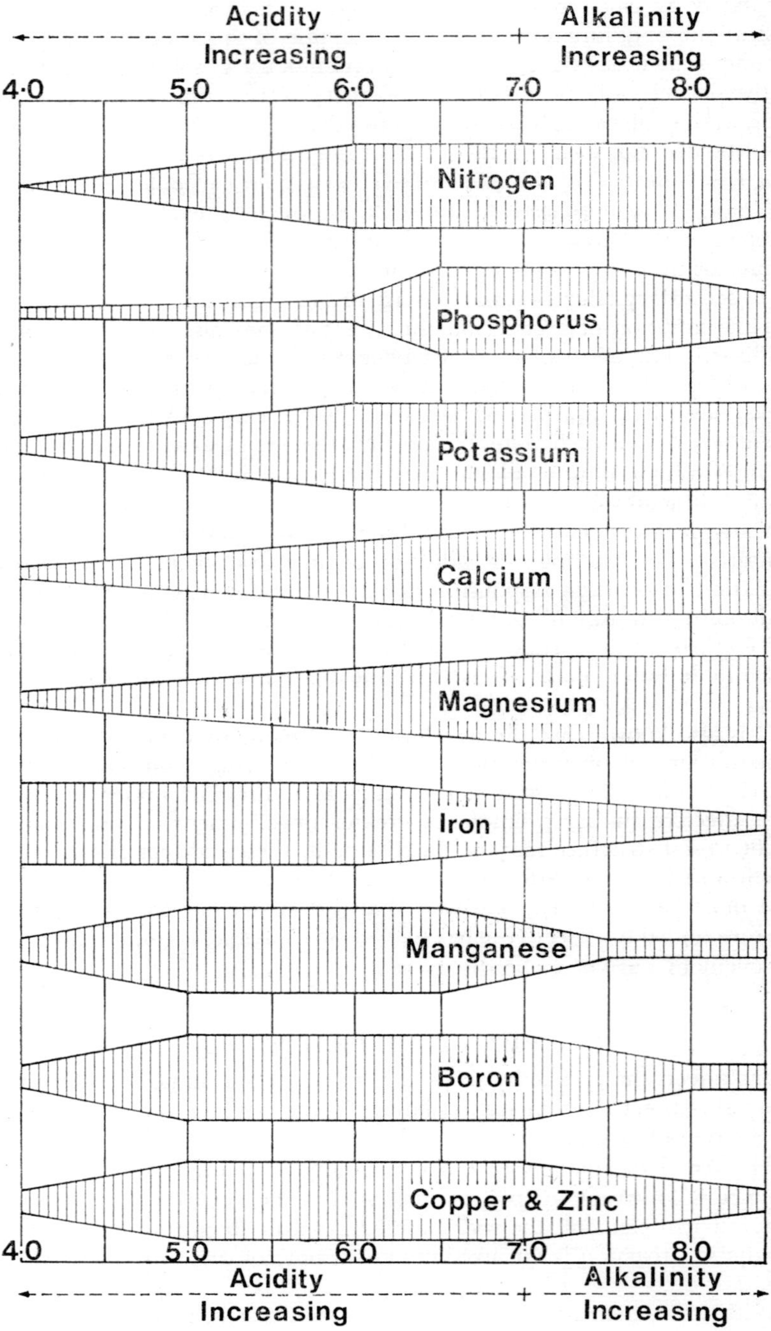

Fig. 1. Influence of soil reaction on plant nutrient availability

toxic when acid conditions are associated with poor aeration—iron, for instance, which is converted to soluble and toxic ferrous compounds. On the other hand, molybdenum becomes more available under alkaline conditions, while phosphates are not only converted into less soluble forms, mainly of iron and aluminium, with increasing acidity, but also at the other extreme with increasing alkalinity when calcium phosphate may be formed.

The supply of calcium and magnesium is more directly affected by liming because of the presence of those nutrients in the liming material. With nitrogen, however, the effect of liming is mainly indirect, and changes in soil acidity affect the activity of the soil organisms, which convert complex nitrogen compounds into the simpler forms used by plants.

On the whole, therefore, the most favourable conditions for the supply of all the plant nutrients is reached at a pH of about 6·5—just below neutral point.

Although the reduction of acidity is usually desirable, there are crops that thrive in acid soils because of a special need for elements that become more available to them under acid soil conditions. In such cases the reduction of soil acidity by liming results in decreased solubility of nutrients, such as manganese, iron and boron, until a point may be reached where the supply is insufficient for the crop and signs of nutrient deficiency appear. Thus on calcareous or over-limed soils not well supplied with trace elements, fruit trees may suffer from a yellowing or chlorosis due to iron deficiency, oats from the Grey Speck disease caused by lack of available manganese, and sugar beet from Heart Rot due to lack of boron.

The special effects of soil acidity and liming on each of the plant nutrients are dealt with more fully in the following sections.

Nitrogen

Soil acidity affects the supply of available nitrogen from resources in the soil, and from added manures and fertilizers. Nitrogen from organic matter, either from applied organic nitrogenous manures or from plant residues, is converted into forms available to plants by the activity of micro-organisms, principally bacteria. Ammonia-producing organisms in the soil are not greatly affected by soil acidity. The organisms responsible for this change show a marked tolerance to acidity and the ammonia formed also helps in the partial neutralization of acidity. On the other hand, the bacteria responsible for the conversion of ammonia to nitrates cannot work efficiently in acid soils, and ammonium salts applied as mineral fertilizers remain near the surface and are only slowly converted into nitrates. The application of lime stimulates nitrification with the production of readily available nitrogen. The most favourable reaction range for nitrification lies between pH 6 and 8.

The process of fixation of atmospheric nitrogen by the bacteria in the roots of leguminous plants is greatly retarded both under strongly acid and strongly alkaline conditions. Not only is the rate of fixation retarded but the conditions for conversion of the organic nitrogen formed are less favourable. A pH of about 6·5 is generally considered most favourable for nitrogen fixation, but a few leguminous plants will grow and fix nitrogen satisfactorily under much more extreme conditions.

Phosphorus

The most favourable condition for phosphate availability is when the pH lies between 6·5 and 7·5. In soils well supplied with exchangeable lime, the water-soluble phosphates of superphosphate are rapidly converted into less soluble calcium phosphates, in which form they are still sufficiently soluble to be available to plants for many years. The complex phosphates in high-soluble basic slag behave similarly, but the comparatively insoluble phosphates in mineral phosphate are of little value to plants in neutral or heavily limed soils. However, on acid soils, phosphates from mineral phosphate can prove as effective, particularly in high rainfall areas, as those derived from the more soluble forms.

Below pH 6·5 as acidity increases, phosphate availability in the soil decreases rapidly. Sooner or later on acid soils, all added phosphates are converted into poorly available forms due to the formation of insoluble iron and aluminium phosphates unavailable to crops. Above a pH of 7·5, that is to say on over-limed soils and those which naturally contain excess of lime, soil phosphates also become less available, because the soluble phosphates are converted into a less soluble calcium phosphate, which after a time is converted into another insoluble and unavailable form.

Potassium

The supply of exchangeable potassium in the soil is often low in acid soils, due to the formation of soluble potassium salts with soil acids and their loss by leaching from the soil. The availability of potassium also begins to fall below a pH of 6·0.

Calcium and Magnesium

The supply of these two nutrient elements, for the same reasons as with potassium, tends to be low in acid soils and less available to plants. This decrease in availability starts as soon as the soil pH falls below the neutral point. Since agricultural limes contain calcium and sometimes magnesium as main constituents, the soil status of calcium and magnesium is also directly affected by liming.

Manganese

Manganese occurs in several states of chemical combination in soils, depending largely on the degree of acidity and aeration of the soil. In acid soils, the availability of soil manganese increases rather rapidly when the pH falls below 6·5, and under very acid conditions may reach toxic levels. At pH values above 6·5, the amount of soluble manganese available for crop growth falls rapidly and, under alkaline conditions, soon becomes a limiting factor in crop growth on some soils, especially with such susceptible crops as oats and sugar beet (see page 13).

Iron

The availability of iron increases in much the same way as with manganese in acid soils, and iron toxicity due to excess of ferrous iron occurs under conditions of poor aeration from impeded drainage. Deficiency of iron (lime-induced chlorosis) is frequently serious in plants growing on thin soils over chalk, and on other soils containing large amounts of lime.

Boron

Excess lime in the soil leads to boron deficiency in susceptible crops such as sugar beet and swedes. In some soils, the content of available boron is so low that even a normal dressing of lime may cause boron deficiency and thus reduce crop yield. When this occurs, the application of borax to the soil will remedy the condition and at the same time make it possible to meet the lime requirement of crops which need neutral or alkaline conditions for normal growth. Boron is needed only in very small amounts and excess leads to boron toxicity, especially in very susceptible crops such as potatoes. For this reason, boron should only be applied when there is a proved deficiency, and then only in the smallest amounts.

Copper and Zinc

Copper and zinc are required by crops in minute quantities, but cases have occasionally been reported of zinc and copper deficiency brought about in susceptible crops by over-liming. Deficiency of these elements may also occur on acid soils due to leaching.

Molybdenum

In contrast with the other minor elements, molybdenum becomes more available to crops as the lime content of the soil increases. Molybdenum deficiency tends to occur on acid soils, and in suitable cases may be controlled by the addition of lime. A more effective measure is to apply both lime and molybdenum. Deficiency is most easily recognized in cauliflower, where the leaves are so reduced and distorted that they consist of little more than the midrib with small pieces of leaf blade attached—a condition known as 'Whiptail'. A similar condition sometimes occurs on soils containing adequate lime, but it is not yet known if it is caused by molybdenum deficiency.

On soils with a high natural content of lime, such as the Lower Lias clays, some clovers and grasses take up abnormally large amounts of molybdenum, and this causes sudden onsets of severe scouring in cattle due to molybdenum poisoning. Land on which this trouble occurs is known as 'teart' and is common in parts of Somerset and Warwickshire. Farmers in affected areas should seek expert advice before liming their pastures.

EFFECTS ON CROPS AND LIVESTOCK

From the foregoing sections it will have become evident that the response of crops to acid soils and to liming are often the outcome of a complex of causes, in which soil pH, calcium status and the many direct and indirect effects of lime on soil activities and transformations all play their part. Apart from the special effects with particular nutrient elements, crops show marked differences in their requirements for lime and their tolerance of shortage.

In general, leguminous crops—especially lucerne, red clover and sainfoin—make the heaviest demands and are among the first to fail as lime shortage develops, but lupins and a few others are exceptions. Sugar beet and barley and, among the grasses, timothy and crested dogstail, also require ample supplies. Some crops, such as potatoes, rye and oats are more tolerant of acid conditions and often form the basis of rotations on acid soils that receive little or no lime.

The figures in Table 3 indicate the pH above which some of the common farm crops will usually grow well on mineral soils, but these figures are only a guide, and many other factors have to be taken into account.

Although crops show different degrees of tolerance to acid soil conditions, yet they are all adversely affected when the soil becomes too acid. This adverse effect is progressive and at first there may be nothing more than a small decline in yield without obvious visual signs. When, however, soil acidity becomes more extreme it is sometimes quite easy to see the signs in the field.

TABLE 3

Tolerance of farm crops to soil acidity

Crop	*pH**	*Crop*	*pH**
Barley	6·5	Wheat	5·5
Lucerne	6·5	Wild white clover	5·5
Sugar beet	6·5	Oats	5·0
Red clover	6·0	Potatoes	5·0
Beans	5·5	Ryegrass	5·0
Peas	5·5	Lupins	4·5
Turnip and swede	5·5	Rye	4·5

*The crop usually grows well at a pH *above* that shown

Arable Land

On arable land, the very acid-sensitive crops, such as lucerne, red clover, barley and sugar beet, may fail or make very poor weedy stands, since they are unable to compete with acid-tolerant weeds like corn spurrey, sheep's sorrel and corn marigold.

Patchiness of crops is a characteristic feature of acid soil conditions, and may often begin to appear at an early stage through the dying out of young seedlings. Affected crops are unthrifty, pale green or yellow, sometimes tinged with dull red or purple and are less capable of withstanding pests and diseases, waterlogging and drought. There may also be characteristic changes in the habit of the plant: cereals, for example, fail to tiller and they produce thin upright flags which develop long wiry tips.

When lime is applied to the soil, some time elapses before it exerts a uniform effect throughout the soil mass and before a steady pH value is attained. The time required depends upon soil type, moisture conditions and effectiveness of cultivations; during this time the soil is a mixture of pockets with excess lime and pockets with insufficient lime, and uniform application of the lime is necessary to avoid these being too large. During unfavourable growing conditions such as drought some of the effects previously described are likely to affect crops at this time, especially on light sandy soils low in organic matter.

While crops generally grow best at pH values above those given in Table 3, other considerations make it desirable to aim for particular pH values for certain crops. Thus very high pH values, especially in light sandy soils low in organic matter, can increase the incidence of scab on potatoes; high pH values in most soils can adversely affect the uptake of magnesium and some trace elements. Low pH values can encourage club root in susceptible crops.

With many soils used for arable cropping there is an inevitable fall in lime status between applications of lime and it is desirable to apply maintenance dressings at a suitable point in the rotation. This is most important when crops like sugar beet, lucerne and barley on one hand, and crops like potatoes, oats and all grass leys on the other hand, are grown in any one rotation. The obvious time to lime is before sowing the former crops or immediately after harvesting the latter crops.

At the present time, rotations are being lengthened by runs of consecutive cereals crops, usually barley, and this helps to avoid liming too close to crops sensitive to excess lime; at the same time, long runs of cereals before a break crop could lead to liming at too infrequent intervals, with consequent slow build-up of soil acidity and large abrupt pH changes when lime is applied. With most arable rotations, and especially on light sandy soils there is much to be said for carrying out maintenance liming at regular frequent intervals.

Temporary Grassland

The lime status of soils under leys of one to three or four years' duration in an arable rotation will in general be governed by the lime requirement of the arable crops; in this case it is preferable that pH values of the soil be maintrined at about 6·5. Care should be taken to ensure that soils sown to all grass leys can provide adequate magnesium for uptake into the herbage. Leys of longer duration should be limed at three to five year intervals to maintain a pH value not less than 6·0.

Permanent Grassland

Soils under permanent grass are best maintained at a pH value of about 6·0; at this level all the commonly sown grasses and white clover yield as well as at 6·5, and the uptake of minerals essential for animal health is better balanced. The lime requirement value determined by soil analysis to bring the pH value of a soil continuing under grass to 6·0 may in practice be reduced depending on soil type.

Under acid conditions, a mat of decomposed roots and dead herbage is frequently found on permanent grass. Clover makes very poor growth or disappears completely, while the better grass species die out and are replaced by a poor wiry type of herbage consisting mainly of bent grass and characteristic weeds. On grassland that has become acutely acid, the vegetation is largely composed of unpalatable species and grazing is uneven. Even where liming produces no very obvious changes, it often improves the palatability and nutritive value of the herbage. On more acid soils, liming results in changes in the herbage constituents that bring about a general improvement in the quality and quantity of the grazing.

Grazing Land

Grazing animals depend on herbage for their supplies of essential minerals. Calcium is particularly important for young stock, milking cows and ewes. On poor, lime-deficient pasture, livestock may be unable to derive sufficient calcium for their well-being from the herbage, which may contain only a quarter as much calcium as is present in a properly limed and good quality sward (see section 'Lime for Livestock' on page 44).

Adequate supplies of lime become particularly important when the whole of the produce is removed from the land. When, for example, grass is removed as hay, dried grass or silage, the lime absorbed by the herbage is no longer returned to the soil through the animal, and the loss of lime from the land is greater and needs to be made good (see Table 1).

Under some special conditions, however, liming may have harmful effects on livestock, and in particular on the 'teart' lands it may lead to severe scouring in cattle as a result of the increased uptake of molybdenum by the herbage (see pages 15 and 44).

Liming Materials

LIMESTONES are widely distributed throughout Great Britain (see Fig. 2). They are sedimentary rocks of different geological ages, which may have been laid down in one of the following ways: (i) by the accumulation of shells or skeletal remains of animals and plants; (ii) by the formation of masses by marine calcareous algae; (iii) by precipitation of the carbonates of lime or magnesia from water; (iv) by the aggregation of small fragments derived from pre-existing limestones.

The deposits laid down originally on the bed of the sea or in lakes were soft and were later subjected to various processes of hardening. The more remote the period at which the deposition took place, the longer the processes of hardening had to act and consequently the limestones of the older geological formations are usually harder than those of more recent origin.

Chalk is composed partly of minute skeletal structures and was laid down under warm marine conditions in more recent times than limestone. It consists of almost pure calcium carbonate, but may contain numerous flints. It varies in hardness, and the softer types are commonly worked for agricultural use.

FORMS OF LIME

In agriculture, the word 'lime' is commonly used to mean not only calcium oxide (quicklime, CaO) but also calcium hydroxide (slaked lime, $Ca(OH)_2$) and calcium carbonate (limestone and chalk, $CaCO_3$), whether singly or in mixture. It also covers magnesian or dolomitic limestones, and burnt limes in which the calcium is associated with magnesium, which is at least as valuable as calcium in correcting soil acidity. In addition, other calcium compounds present in soils, crops and feedingstuffs are also sometimes spoken of loosely as 'lime', though the word is then used mainly as an equivalent for calcium.

The sources of the materials specifically produced for agricultural liming are outlined in the diagram. Lime products that are forms of calcium oxide (CaO), calcium hydroxide ($Ca(OH)_2$) and calcium carbonate are included in the diagram. All products not specially labelled CaO or $Ca(OH)_2$ in the diagram are forms of calcium carbonate ($CaCO_3$). The corresponding magnesian limestones and limes ($MgCO_3$, $Mg(OH)_2$ and MgO) are indicated by an asterisk (*).

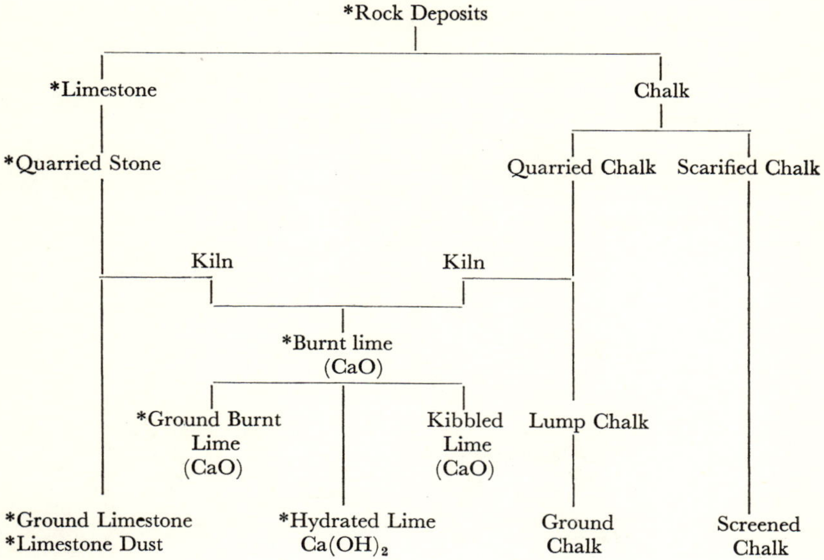

Limestone and chalk are used in many industries as well as agriculture, for example, iron and steel, chemical processes, road-making and cement manufacture.

Calcium Carbonate

In comparison with quicklime, about twice as much ground limestone or chalk has to be applied to give an equivalent rate of dressing. Limestone and chalk cost much less per ton but the larger quantity involves higher transport charges unless these materials can be obtained nearer to hand. Dressings of carbonate, even at heavy rates, cause no scorching of crops and top dressings can be safely given, if necessary. It is not harmful in large doses and may be applied as a top dressing to growing grass or young crops.

Ground Limestone and Ground Chalk

These are natural limestone or chalk ground to a fine powder. A good sample is almost pure calcium carbonate and is equivalent to about 53 per cent its weight of calcium oxide. Calcium carbonate in this form can be stored for an indefinite period without fear of deterioration of the material or destruction of the bags. It is much less unpleasant to handle than quicklime, and is very suitable for mechanical spreading.

The quality of ground limestone depends on the quality of the limestone used, and on the fineness of grinding. There is a limit to the fineness of grinding, beyond which the cost makes the price prohibitive. Furthermore, very fine limestone flour is difficult to distribute, either by hand or machine. The Statutory Statement given to the buyer (p. 27 and 47) shows, in addition to the Neutralizing Value (see page 27), the percentage of the material which will pass through a prescribed sieve (British Standard Test Sieve, mesh number 100). A product with 40 per cent passing this sieve is regarded as a good standard for the normal rate of dressing. Heavier dressings should be applied when coarser products are used. A reasonably coarse product is

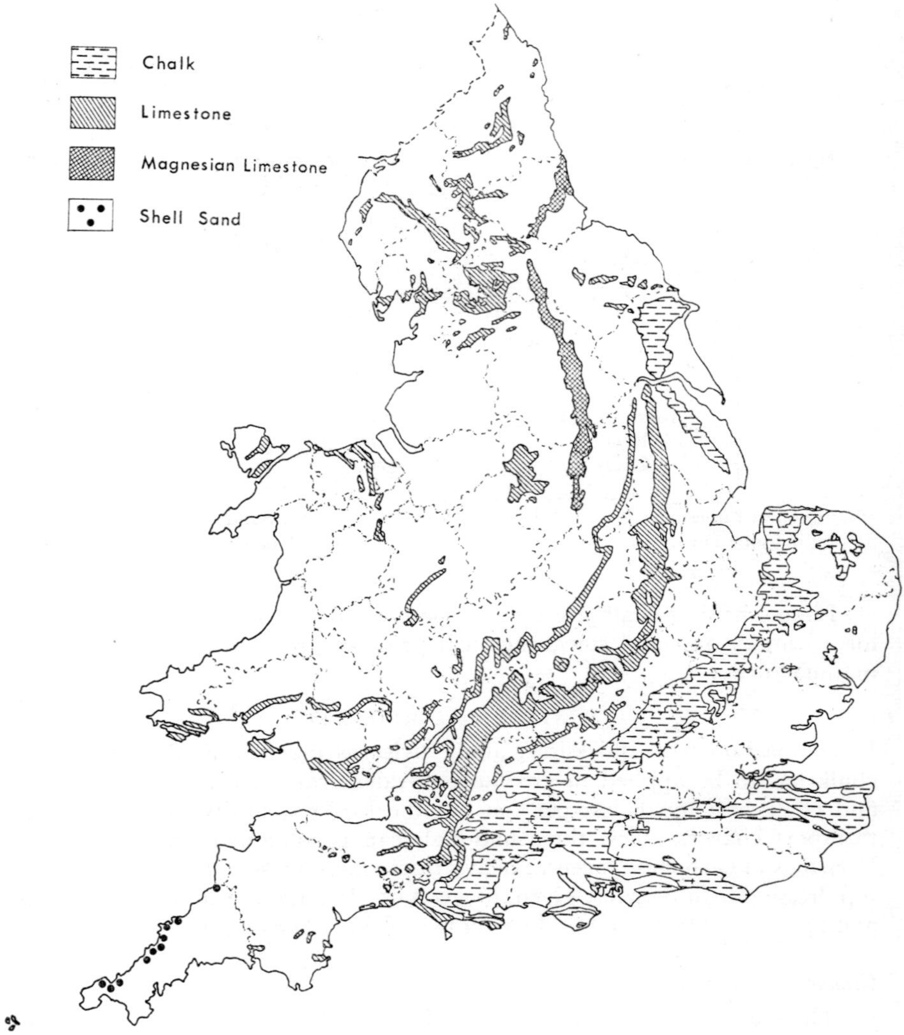

Fig. 2. Map showing the chief limestone areas

cheaper to produce, runs more easily through a drill, and is not so rapidly washed out of the soil.

There are considerable quantities of chalk available in some counties in the east and south-east of England. On account of the high water content of the material as it is quarried, the chalk often has to undergo preliminary drying, which adds appreciably to the cost of production; but the cost of grinding is lower than for limestone. Ground chalk and ground limestone have similar agricultural uses and effects.

Screened Chalk

Beside the screenings resulting from the excavation of lump chalk for the manufacture of burnt lime from chalk, screened chalk is widely produced in the south and east of England. The usual method of manufacture is as follows: an area of soft chalk is bared of overburden and scarified by tractor-drawn harrows. The resulting disintegrated particles are collected together, usually by some form of bucket or bulldozer, and screened. The screens are often static but occasionally vibratory screens are used to produce the finer grades. In general, the products are graded from all passing 2 in. mesh to all passing through a $\frac{1}{4}$ in. mesh. The screened chalks are relatively cheap and pure and are usually sold per ton or per cubic yard. They are most easily produced in dry weather.

Lump Chalk

The application of large quantities of raw chalk (15–20 tons or more per acre) was formerly a general practice in some parts of the country. In recent years, however, this method of applying lime has been largely discontinued, although it is still practised in some districts, especially in the southern counties. The chalk was dug from pits in late autumn and spread directly over the land. No attempt was made to break up the lumps, winter frosts and spring cultivations being relied upon to reduce them to a sufficiently fine state.

Calcium Oxide

Caustic limes generate considerable heat when in contact with moisture, and increase in volume on slaking. Consequently, they should not be stored near inflammable materials, such as straw or hay, owing to the possible fire risk, and they should also not be stored in a confined space. Provided the weather conditions are suitable, burnt lime in the field slakes or 'falls' naturally into a fine state of division and some of it passes into solution. It can then be more thoroughly incorporated in the soil, and for this reason may act more rapidly and prove suitable for emergency treatments (see page 31). But if the lime has been overburnt or if slaking is inefficient, this advantage is lost.

Lump Quicklime

This is known as burnt lime, lump lime, lime shells, clot lime, cob lime, quicklime, hot lime or caustic lime, and is obtained by burning chalk or limestone in a lime kiln. Quicklime, when exposed to the atmosphere or applied to the soil, becomes slaked and is then gradually converted back to calcium carbonate.

Good samples of burnt lime should have a Neutralizing Value of not less than 80 expressed as calcium oxide. The highest grade, not always available for agriculture, is sold as 'hand picked' lump lime or 'screened' lime and contains none of the ash from the fuel. Unscreened lime is the material as it comes from the kilns and is usually sold as 'run-of-kiln'. The screenings are sold as 'small lime' (see page 22). Incompletely burnt samples may be detected by breaking some of the larger lumps, within which a hard core of unburnt limestone will be found. Such incompletely burnt lumps do not fall to a fine powder when the lime is slaked.

The usual method of applying lump lime is to put it on the land in small heaps and allow it to slake naturally. It may then be spread with a shovel and harrowed in at once. This method can only be practised efficiently when at least two tons per acre are applied. Moreover, it is open to the objection that, especially in rainy weather, too much moisture is apt to be absorbed by part of the heap, with the result that the lime becomes pasty and afterwards dries to form hard lumps of calcium carbonate. On the other hand, in dry and frosty weather, lump lime may slake so slowly that it carbonates before it powders completely and may form hard nuggets of carbonate which are of little value. To overcome this difficulty, the heaped lime may be covered with soil, and after a period it will fall to a fine dust.

The rapidity with which a sour soil responds to an application of burnt lime depends very largely on the care that has been taken to ensure uniform distribution. Spreading lump lime direct from carts, without first allowing it to slake, is only too frequently the cause of poor results; the largest lumps can never be efficiently spread, no matter how careful the cultivation.

Kibbled Lime

In order to facilitate efficient distribution and mechanical spreading, many lime burners now offer kibbled lime for use on farms. This material is burnt lime which has been crushed into small pieces usually not greater than 1 in. cube. Machines are available which will spread this type of product at the rate of about 1–1½ tons per acre. Contract spreading charges are generally reasonable. Kibbled lime is slightly dearer than burnt lime but cheaper than ground lime.

Ground Quicklime

This consists of burnt lime that has been crushed so that it will pass easily through a manure distributor. It is a convenient form of quicklime for use when the rate of application is less than 2 tons per acre, but the supply to farms is limited by reason of industrial demands. The ground lime slakes in the soil to a still finer powder and thus ensures excellent distribution. On account of the cost of crushing and bagging, ground lime is usually considerably dearer than lump lime, but the ease with which it can be handled and distributed makes it a very useful form of quicklime.

Ground lime quickly absorbs moisture, swells, and may burst the bags in which it is supplied if they are filled too full. Bags should be laid on their sides in store and not kept for more than a week or two. Quicklime must be spread without delay, which may be inconvenient if the weather is unfavourable. The most serious disadvantage of ground lime is that it is exceedingly unpleasant to handle on account of its caustic nature.

Small Lime

The small lime remaining after the selection of the best lump lime is often available for agricultural purposes at a cheap rate, and, provided that it does not contain too high a proportion of cinders and waste, it is quite a suitable form of quicklime.

Calcium Hydroxide

Hydrated Lime (or slaked lime)

This is often supplied under various trade names. It is usually very pure and in a fine state of division. It has the advantage over quicklime of being more stable, it does not burst the bags, and may be conveniently stored until the weather is suitable for effective distribution. It is, however, much more expensive per unit of Neutralizing Value than other forms. Moreover, because of the heavy demand for hydrated lime in the industrial market, supplies to agriculture are very restricted. For ordinary agricultural use, therefore, this form of lime is generally too dear if others can be obtained near at hand. Even horticultural users who traditionally favour hydrated lime would be wise to consider using some alternative material such as ground limestone or ground chalk.

Magnesian Lime

Ground Magnesian Limestone

Magnesian limestone occurs as a valuable liming material mainly in the north and midlands. It has additional special uses, on account of its magnesium content, for correcting magnesium deficiency in soils. When soils are acid, dressings of magnesian lime will help to correct both soil acidity and any magnesium deficiency. Magnesian limestones are a mixture of magnesium carbonate and calcium carbonate, although the proportion of magnesium carbonate varies. By law, the magnesium content must be not less than 3 per cent (or approximately 10 per cent $Mg\,CO_3$). Those with a high magnesium carbonate content are sometimes known as dolomite.

Magnesian limestones have a slightly higher Neutralizing Value than the calcium forms, but in practice they are about equivalent to ground limestone and chalk as liming materials.

A product ground so that 40 per cent will pass through a 100 mesh sieve is generally desirable. When coarser grades are used, heavier rates of application are necessary.

In Yorkshire, large quantities of magnesian limestone dust are available for use as a liming material and this has been found to be nearly as effective as the more finely ground limestone. In wet areas, dust, of which 25 per cent passes through the 100 mesh sieve and all passes through the $\frac{1}{8}$ in. sieve, is only a little less efficient than the more finely ground product.

Burnt Magnesian Lime

When magnesian limestone is burnt in a lime kiln the calcium and magnesium carbonates are converted into the corresponding oxides. Although the burnt product can be carted out into small field heaps in the same way as ordinary burnt lime, it is safer to allow it to slake first. Burnt magnesian lime slakes rather slowly and it is better to slake it in a heap at the entrance to the field. Alternatively, the burnt product may be purchased in ground form which can be spread mechanically.

Weight for weight, burnt magnesian lime is more effective than magnesian limestone in neutralizing soil acidity; as a rough guide 1 ton burnt magnesian lime can be taken as equivalent to 2 tons ground magnesian limestone. Despite the fact that it is necessary to use twice as much ground limestone as burnt lime, the former is always cheaper, unless transport costs are high.

Prejudice against Magnesian Lime. In some parts of the country there is a long-standing prejudice against the use of burnt magnesian lime, dating from the times when it was customary to apply dressings of 5 tons or more per acre. It is found that when *heavy* dressings of the *burnt* material are given, crop yields are reduced, although this may not last longer than a season or so.

Even more pronounced and long-lasting harmful effects may be seen on the sites of heaps where burnt magnesian lime has slaked, no doubt due to its strongly caustic nature.

When lighter dressings of burnt magnesian lime are given—say, not exceeding 2 tons per acre—there is no reason to fear any reduction in crop yield, particularly if slaking in heaps is avoided. In any case there need be no hesitation at all about liming with ground magnesian limestone and magnesian limestone dust, which are both perfectly safe.

Lime Mixtures

Ground Mixed Lime

This form of lime is frequently a mixture of calcium oxide, usually a fairly small proportion, and calcium carbonate. It is often produced by grinding a mixture of chalk or limestone and burnt lump lime. The product in most cases is not as fine as ground limestone, but the lime content is higher owing to the inclusion of calcium oxide. As compared with ground limestone, the content of quickly available burnt lime in the mixture will tend to offset its inferior fineness. Its Neutralizing Value and unit cost will indicate its relative usefulness. Sources of supply of mixed lime are concentrated in the midlands and the south of England.

By-Product and Waste Limes

Waste lime products are not scheduled under the Fertilisers and Feeding Stuffs Act. Many of them are useful and cheap materials suitable for mechanical spreading, but with others there may be disadvantages. Although under present arrangements farmers normally take up materials from safe sources only, advice is available and should certainly be taken before attempting to use any form of waste lime from unknown or previously untried sources. Disadvantages which may be found with some forms of waste lime are mentioned under the particular materials below, but the main points to be noted may be summarized as follows:

The material may contain substances injurious to plants, necessitating spreading some weeks before crops are sown.

The material may be in a wet and sticky condition, which makes even distribution on the land very difficult.

Because of high moisture or bulk, waste limes are not worth transporting long distances, and can only be used economically near the works that produce them.

Neutralizing Value may not be uniform, and it may not be easy to assess the economic value of the material at first glance.

Hydrate Rejects or Hydrator Tailings

This is a by-product of the manufacture of hydrated lime. It consists in the main of unslaked lime, slaked lime, unburnt stone and impurities in the original burnt lime.

It is relatively cheap, and when available in a fresh condition, it is a useful product for farms near to the works. If, however, tailings are stored in the open for any length of time, much of the oxide and hydrate is converted to calcium carbonate, and rain-water may also be absorbed. Consequently, the material becomes of less value per ton and may be difficult to spread evenly.

Sugar Beet Factory Lime Sludge

A large quantity of waste carbonate of lime is produced at sugar beet factories. It usually contains more than half its weight of water, and about

30 to 40 per cent of calcium carbonate, but the state of the material varies with the amount of water present. When it is partially dried, about 2 tons may be regarded as equivalent to 1 ton of ground carbonate. Lime sludge is disposed of at a nominal price, and, within short distances of the factory, it provides a convenient source of lime. On account of its moisture content and the presence of some organic matter, the sludge should be spread and harrowed in some weeks before seeds are to be sown. When artificially dried the product is comparable with other forms of dried carbonate of lime.

Sulphate of Ammonia Waste

In the north of England very large tonnages of a precipitated carbonate of lime are available as a by-product from the production of sulphate of ammonia. The by-product has a Neutralizing Value of at least 30 and contains about 20 per cent moisture and 2 per cent sulphate of ammonia. The material is fairly easy to spread and is a good liming agent particularly for grass. The nitrogen content of the material should be taken into account when planning subsequent fertilizer applications.

Sulphuric Acid Waste

A waste lime greyish-green in colour, from the production of caustic soda and other chemicals, is available in various parts of the country. Most of this material has been exposed to the weather for many years and contains only small amounts of sulphides and chlorides. The moisture content varies, and the material from some dumps needs opening up to wind and sun before it can be easily applied to the land. Air-dried samples usually have a Neutralizing Value of about 30.

Tannery Waste Lime

This is obtained in small quantities, but may contain sulphides that are injurious to plants. It should be exposed to the air for some months before coming into contact with a growing crop. The waste lime from some tanneries may contain arsenic, and such lime should never be used on the land.

Tar Distillery Waste

A waste lime from production is available at some tar distillation plants. It may contain up to 50 per cent water, but in some instances is partially dried, with a water content as low as 30 per cent, and it then has a Neutralizing Value of about 30.

The waste lime may contain small amounts of phenols and as a safeguard should be applied to the land in autumn or winter.

Waterworks Waste

This by-product from waterworks consists mainly of calcium carbonate precipitated from the water, although a small proportion of hydrated lime may be present. It usually contains a large amount of moisture, but when dry it falls to a fine powder. Left in heaps, and partially dried, it can be spread by means of shovels. A few waterworks have installed drying plants where this waste is partially dried. This makes excellent material even though its moisture content is still high (25 to 30 per cent). It can be used at about double the rate for ground carbonate.

Whiting Sand

This is a waste product from the manufacture of whiting. It is a pure form of carbonate of lime, but as the chalk is ground under water, and the waste material stacked in the open, it usually contains a good deal of moisture. When dry it falls to a fine powder. During summer and autumn the material from old heaps is usually dry enough to be spread with ease.

Miscellaneous Wastes

Many industrial processes employed in soap works, paper works and bleach works yield calcium carbonate in a very fine state of division, containing 30 to 40 per cent of water, often with small quantities of caustic lime and soda. These wastes can generally be obtained at nominal prices, and are usually spread by contractors. When spread on the land they can dry rapidly and fall to a fine powder. Provided there are not more than minor traces of caustic alkali, they are not likely to cause trouble. The material should be applied some time before the crops are to be sown, or it can be used as a grassland dressing. About 2 tons of wet material equal 1 ton of ground limestone or chalk.

Other Sources of Lime

Calcareous Marl

The long established practice of marling consists of applying to the land an intimate mixture of small chalk and clay. There are, however, natural deposits containing finely divided calcium carbonate, mainly in north Wales and the north of Scotland. The materials are often wet and difficult to handle, but when air dried are fairly cheap and efficient sources of lime for local use.

Limestone Dust

This material has been on the market in plentiful supply and at low prices for some years, and its usefulness depends entirely on the proportion of finely powdered limestone it contains. Considerable quantities of this dust arise during the manufacture of roadstone, chippings, etc., and when delivered direct from the hopper the dust is dry and often contains a considerable proportion of fine material. In the better limestone dusts produced in this manner as much as 35 per cent of the material passes the 100 mesh sieve, and generally all the material passes $\frac{1}{8}$ or $\frac{1}{16}$ in. sieve. In some north-eastern districts, certain beds of magnesian limestone contain 'pockets' of fine limestone dust which must be removed when the rock is quarried. This dust contains a high proportion of magnesium carbonate.

The unscreened product may contain a considerable proportion of coarse material and pieces of rock sufficiently large to prevent its application by means of a farm distributor. With limestone dusts delivered direct from the hopper of roadstone plants, the firm will often guarantee a fineness of 100 per cent passing $\frac{1}{8}$ in. or $\frac{1}{16}$ in. mesh, and the material supplied in this fashion is generally quite dry and easily spread.

The material when loaded direct from heaps on the floor of the quarry is usually dry enough to be spread but it may, through exposure to rain, contain a high proportion of moisture when delivered. To prevent the absorption of additional moisture in transit trucks should be sheeted.

Two tons per acre may be considered a minimum dressing of this material, which can be applied with shovels direct from carts, or from a mechanical lime distributor in which occasional outsize pieces will not injure the mechanism.

Sea Sand

In certain coastal areas, calcareous sea sand is used as a source of lime. Samples having a Neutralizing Value about 30–45 are extensively used near the coast of Cornwall and in the north of Scotland, but the lime content varies considerably.

LIME CONTENT WARRANTY

Under the Fertilisers and Feeding Stuffs Act, 1926, a statutory statement which has effect as a warranty must be given to the purchaser by the seller in respect of the principal liming materials. This statement relates to the quality of the article.

The present regulations (1968) are given in the Appendix (see page 46). The liming materials scheduled in the regulations are listed and defined, and details are given of the particulars required in the statutory statement for each article. If any article is sold under a name listed in the Fourth Schedule, a warranty is thereby given that the article conforms with the definition in the Schedule (see the second column of the table in the Appendix). The lime content of a material must now be expressed in terms of a figure for *Neutralizing Value*, instead of as the percentage content of the lime compounds present and the percentage-equivalent as calcium oxide (see explanatory information on page 28).

Every purchaser of an article, in respect of which a statutory (or voluntary) statement has been given, has the right under the Act to have a sample taken for analysis and to be furnished with a certificate of the result of the analysis. The certificate will show him whether or not the composition of the article complies with or differs from the warranty.

All that the farmer or other purchaser has to do is to request an Official Sampler appointed by the County Council or County Borough Council, as the case may be, to take a sample of the article. A fee may be charged for this service. The sample must be taken within fourteen days of delivery of the liming material or of receipt of the statutory statement, whichever is the later. If the analysis discloses a deficiency or other breach of warranty, it is for the purchaser to decide whether to make a claim against the seller or not. No prosecution can be instituted by a Local Authority against a seller in respect of any deficiency where the sample has been taken on a farm by an Official Sampler. The name and address of the nearest Official Sampler can be obtained on application to the Clerk to the Local Authority.

NEUTRALIZING VALUE

Neutralizing Value is an expression of the value of a liming material for neutralizing soil acidity. It is determined by laboratory tests and is expressed as calcium oxide (CaO). For example, the statement that burnt lime has a Neutralizing Value of 90 signifies that 100 lb of it has the same effect in neutralizing acidity as 90 lb of calcium oxide (CaO).

Since the main purpose for which lime is used is to neutralize soil acidity, the Neutralizing Value is the most important property by which the various liming materials may be compared. Under the old regulations (now revoked), the amount of calcium oxide, calcium carbonate or hydroxide in the material and the equivalent of calcium oxide (expressed as percentage of CaO) had to be stated. This took no account of other neutralizing agents, particularly magnesium oxide or carbonate, that the material might contain, although weight for weight magnesian limes have a greater neutralizing power than the calcium forms. The Neutralizing Value is, therefore, a more satisfactory figure from the farmer's point of view.

The figure for Neutralizing Value (N.V.) of a calcium liming material is usually about the same as the old figure for 'percentage CaO' (or CaO equivalent) but whenever magnesian lime is present in a material the N.V. will be higher than the old figure and may even be over 100 in a lime of high magnesium content. For example, a burnt magnesian lime containing 50 per cent calcium oxide (CaO) and 40 per cent magnesium oxide would have a Neutralizing Value of about 106, that is to say 100 lb of this material would equal 106 lb of calcium oxide in its effectiveness for neutralizing soil acidity.

Calculation of Neutralizing Value

If a figure for N.V. is not given, an approximate value can be calculated if the percentage contents of calcium and magnesium oxides (CaO and MgO) and carbonates ($CaCO_3$ and $MgCO_3$) are known for the material. Each percentage is first converted into the equivalent percentage of calcium oxide by multiplying it by one of the following conversion factors. The sum of the calcium oxide percentages is approximately the Neutralizing Value.

Conversion Factors

MgO to CaO	1·4
$MgCO_3$ to $CaCO_3$	1·2
$CaCO_3$ to CaO	0·56
$MgCO_3$ to CaO	0·67

Example A

A burnt lime contains 75 per cent CaO and 10 per cent MgO

Convert to CaO:

75 per cent CaO $=75$

10 per cent MgO $=10\times1\cdot4=14$

Therefore *Neutralizing Value* is 89

Example B

A limestone contains 65 per cent $CaCO_3$ and 25 per cent $MgCO_3$

(i) Convert to $CaCO_3$

65 per cent $CaCO_3$ $=65$

25 per cent $MgCO_3$ $=25\times1\cdot2=30$

$CaCO_3$ Equivalent $=95$

(ii) Convert $CaCO_3$ to CaO:

95 $CaCO_3=95\times0\cdot56=53\cdot2$

Therefore *Neutralizing Value* is 53

Alternatively, convert direct to CaO:

65 $CaCO_3=65\times0\cdot56=36\cdot4$

25 $MgCO_3=25\times0\cdot67=16\cdot8$

CaO Equivalent $=53\cdot2$

Therefore *Neutralizing Value* is 53

COMPARING COSTS

In comparing the price of one liming material with another, it is necessary to take into account the Neutralizing Value of the products, their costs per ton (including the cost of delivery) and the cost of spreading on the land. The point is that a low grade, bulky product may appear cheap on first cost, but what really matters to the buyer is the Neutralizing Value of the material and its whole cost on the land. With bulky material, the advantage of a low price per ton can soon be lost in low Neutralizing Value and the extra cost of transport and labour in spreading the larger quantity needed. Price comparisons are, therefore, best made on the basis of the cost of a unit of Neutralizing Value (N.V.), including delivery and spreading on the land.

Calculation of Unit Cost

The cost of a unit of Neutralizing Value is obtained as follows:

$$\text{Price per unit} = \frac{\text{Price per ton of the material}}{\text{Neutralizing Value (N.V.)}}$$

The 'price per ton' should include: first cost of material, delivery cost and spreading cost, all per ton.

The figure for N.V. is obtained from the Statutory Statement required under the Fertilisers and Feeding Stuffs Act, or from a voluntary statement of analysis in the case of waste limes and other forms of lime, for which a declaration of N.V. is not compulsory under the Act.

Example 1

A ground burnt lime has a Neutralizing Value of 85 and costs 46*s.* 0*d.* per ton ex works; the cost of delivery and spreading, inclusive, is 10*s.* 0*d.* per ton.

$$\text{The cost of a unit of N.V.} = \frac{46+10}{85} \text{ shillings spread on the land}$$

$$= \frac{56}{85} \times 12 \text{ pence}$$

$$= 7\cdot 9 \text{ (say } 8d. \text{ per unit)}$$

Using the same method of calculation its cost ex works would be found to be 6½*d.* per unit.

Example 2

A ground chalk has a Neutralizing Value of 53 and costs 24*s.* 0*d.* per ton ex works; the cost of delivery and spreading per ton is 10*s.* 0*d.*

$$\text{The cost of a unit of N.V.} = \frac{24+10}{53} \text{ shillings spread on the land}$$

$$= \frac{34}{53} \times 12 \text{ pence}$$

$$= 7\cdot 7 \text{ (say } 7\tfrac{3}{4}d. \text{ per unit)}$$

Its cost ex works would be 5½*d.* per unit.

If the distance of haulage were greater, and the cost for delivery and spreading increased to 20*s.* 0*d.* per ton, the cost per unit spread would be 9½*d.* for burnt lime, but 10*d.* for the ground chalk. The ground chalk is thus cheaper ex works than burnt lime, but the same price with 10*s*. 0*d*. haulage costs, and dearer with 20*s*. 0*d*. haulage costs.

Example 3

A ground burnt magnesian lime has a Neutralizing Value of 96 and costs 48*s*. per ton ex works; the cost of delivery and spreading is 10*s*.

$$\text{The cost of a unit of N.V.} = \frac{48+10}{96} \text{ shillings spread on the land}$$
$$= \frac{58}{96} \times 12 \text{ pence}$$
$$= 7\tfrac{1}{4}d.$$

The cost ex works would be 6*d*. per unit. The cost per unit of this burnt magnesian lime is less than that of the burnt lime in Example 1.

Table 4 gives other examples for a range of materials priced under the same three sets of conditions. It illustrates the point that bulky products of low grade lime content (low N.V.) are often cheapest per unit at the works, but haulage and spreading costs increase the unit price more than for products of high lime content (high N.V.). Long distance haulage of bulky products soon makes the price look unattractive. The cost of bags and bagging are also unfavourable to the unit price of ground limestone and chalk in comparison with burnt lime, and the advantage to unit price in buying limestone and chalk loose on short distance haulage is brought out clearly.

TABLE 4

Examples of unit costs of liming materials (including delivery and spreading)

Liming material	Neutralizing value*	Price ex-works*	Unit costs of material: pence		
			At works	Delivery and spreading included*	
				At 10s. per ton	At 20s. per ton
		s. d.			
Lump lime . .	90	85 0	11·3	12·7	14·0
Ground burnt lime .	90	90 0	12·0	13·3	14·7
Ground limestone .	48	25 0	6·3	8·8	11·2
Ground chalk . .	48	30 0	7·5	10·0	12·5
Limestone dust . .	35	12 0	4·1	7·5	11·0
Screened chalk . .	46	18 0	4·7	7·3	9·9

* These figures are examples only to illustrate the effect on unit price of increases in these items.

The Unit = 1% of 1 ton.

It should be emphasized that all these prices are merely illustrations, because the actual prices and N.V. of liming materials vary from district to district on account of the differences in haulage costs and in the materials available. Calculations of unit price should, therefore, be based on local quotations. It should further be realized that there are other considerations to be taken into account besides price, such as ease and convenience of handling, cost of larger dressings necessary with damp or lumpy materials, and so forth. When liming materials are used which are difficult to spread they have to be applied at heavier rates than when finer grades are used, and any small advantage in unit price may be more than offset by this.

LIME SUBSIDY

Under the Agricultural Lime Scheme, purchasers of two or more tons of lime which is used to improve the fertility of agricultural land may apply for a contribution towards the costs they incur in acquiring, transporting and spreading lime on the land.

Members of recognized allotment associations and gardeners' societies may also benefit.

All the usual forms of lime from approved sources are eligible and further information may be obtained from approved suppliers or from any Regional or Divisional Office of the Ministry.

The Practice of Liming

THE reaction of liming materials on soil is slower than that of fertilizers and because of this, the time of year when applications are made need not be so critically controlled. This fact is important in allowing farmers to choose a time when cropping leaves the land free or when dry weather makes possible the movement of heavy transport and spreading equipment over soils which become boggy when wet. It also helps contract services to supply and spread lime reasonably steadily throughout the year, dealing with grassland during the summer, when arable crops occupy much of the land and make access impossible. Farmers are, therefore, advised to place their orders with suppliers or contractors as early as possible and to be ready to accept delivery.

When emergency treatment has to be given, whatever the product used, the importance of a sufficiently finely ground form of lime must be realized; otherwise, the lime will fail to act quickly or uniformly enough to put matters right. It is not enough to ask the supplier merely for *ground* lime, chalk or limestone, which under the present regulations covers a wide range of grades, but the supply of a sufficiently fine grade should be insisted upon.

The action of lime is not instantaneous and application well in advance of sowing time is important. Uniform distribution is also important, and care should be taken to ensure this whether a simple type of spreader, or more elaborate machinery, is being used.

The correct dressing depends on the soil conditions and type of cropping. Once a field has attained a pH of about 6·5, it is most economically kept in condition by regular moderate dressings every few years.

Unless there are special reasons for the use of the oxide (burnt lime) or the hydroxide (hydrated lime) the usual form applied is the carbonate (limestone or chalk). If, however, results are required within a week or two of application, burnt or hydrated lime in a very finely divided state which can be intimately mixed with the soil is preferable (see page 21).

When lime is spread on a field, the speed with which it will reduce the acidity of the soil depends on the degree of acidity and the thoroughness with which the lime is mixed with the soil, how deeply it penetrates, the rate per acre of application and the size of the particles spread. Up to a point, the smaller the lime particles the more quickly and evenly they mix with the soil, but the larger particles are more lasting, and if not too large, form a useful reserve of lime throughout the rotation. The coarser the grade the larger is the dressing needed to achieve the same result. If land is very acid lime should be applied well in advance of cropping, and if an acid-sensitive crop is to be grown liming should preferably be done in the previous year.

The depth of penetration can be controlled by the method of distribution. The method which should generally be used is to apply the lime as a top dressing on the ploughed field, stubbles or on grass. At first, it will affect the immediate surface only but rain gradually washes it further in. When, however, a soil is very acid throughout its depth or the subsoil more acid than the surface soil, it is sometimes advisable to plough in half the lime dressing and spread the other half on the ploughed surface. Ploughing in enables it to become more thoroughly mixed with the soil in the top 6–8 in. This is especially necessary when a sensitive crop such as sugar beet or barley is to be grown, for if only the top inch or so of soil is affected by the lime the growing roots will pass through it into the acid soil below or may even turn aside from the acid layer and run horizontally.

Lime does not move laterally but gradually works downwards in the soil, and so care must be taken to see that the spreading is done as carefully and evenly as possible. Poor work only results in patchy crops.

Materials Used

The various forms of liming materials are fully described in an earlier section (see page 18) but the forms most generally available for agricultural purposes are those obtained from natural deposits of limestone and chalk as carbonate of lime (calcium carbonate), and from the processed raw materials as burnt lime (calcium oxide). There are also the magnesian limes, which are locally available, particularly in the midlands and north of England.

Ground Limestone or Ground Chalk

This is readily available in most parts of the country. The fact that it is not caustic and does not burn the skin facilitates its handling and spreading.

For practical purposes, 2 tons of ground limestone, chalk or dried carbonate of lime, or 3–4 tons of shell sand or of moist waste carbonate of lime are equivalent to about one ton of ground burnt lime. Certain types of carbonate of lime contain some magnesium carbonate, which is slightly more effective than calcium carbonate in neutralizing acidity.

Burnt Lime

Lump lime is heaped in the field to slake and break down to powder under the influence of weather before spreading, but it should not be left long enough to form a slurry or become too wet so that a crust forms on the surface of the heap.

The weathering process converts burnt lime (quicklime or calcium oxide) into slaked lime (calcium hydroxide) by combination with moisture in the air, but if it is left out for any length of time it reverts to carbonate of lime, by combining with carbon dioxide from the air, and sets hard.

Before finely ground lime products became generally available, burnt lime had special uses when a quick action was needed, because of its property of falling to a fine powder on slaking, which under favourable conditions resulted in rapid mixing and early chemical effectiveness in the soil. It is still usable in the pre-ground form for emergency treatments, but very often other finely ground products, such as ground carbonate or the more expensive hydrated lime, are just as effective and safer to use.

Ground burnt lime may be used where there is a considerable mat on grassland intended for ploughing up. Before ploughing, the lime should be well worked into the mat by drastic harrowing. Hydrated lime is a very suitable form to use to save a crop already in the ground and showing signs of suffering from too great acidity, but any finely divided form of lime is suitable for applying to an unlimed field before sowing an acid-sensitive crop.

Magnesian Limestone

In addition to its general usefulness as a liming material in the districts in which it is produced, magnesian lime has special uses for correcting magnesium deficiency under certain conditions on acid soils. Some strongly leached soils, especially those of a sandy nature, are deficient in magnesium as well as calcium. On such soils crops which are susceptible to magnesium deficiency, e.g., potatoes, sugar beet and oats, show characteristic symptoms. Potatoes show an interveinal yellowing in the centre of the leaf followed by scorching, and sugar beet, interveinal yellowing spreading in from the margin. Similar symptoms may arise under quite different conditions and from other causes, and these should be carefully distinguished. When, however, these problems are proved to exist on acid soils it is desirable to give part or all of the lime requirement as high grade ground magnesian limestone, as this supplies magnesium as well as correcting the acidity.

Other special uses of magnesian limestone are dealt with in the section on lime in horticulture (see page 38).

RELATIVE VALUES OF MATERIALS

Although the relative values of liming materials can be readily determined by laboratory tests (see page 27), it does not necessarily follow that these will provide a complete guide to what happens in the field, because the rate of breakdown of different materials in soil may vary, and the fineness of grinding may influence their effect on the soil. Much field experimental work has been done on this point.

In trials of this sort the various kinds of lime are spread by hand on comparatively small areas so that the application is more even and the incorporation more thorough than in normal agricultural practice. Most of these trials were carried out in the north and west of England, where the rainfall is high and crops are generally able to tolerate a higher degree of acidity. Consequently, it is probable that in the drier eastern side of the country larger dressings would have had to be given to obtain similar results, and that fineness of grinding (to smaller than 35 mesh) would have proved to be of greater importance.

Ground Carbonate and Burnt Lime

In two series of experiments, burnt lime was compared with ground limestone, and the effect of grinding limestone to varying degress of fineness was also examined.

In one series on 33 crops, burnt lime was compared with ground limestone of three finenesses of grinding. Sugar beet, mangels and barley all showed large benefits from liming—up to 100 per cent increases over unlimed crops being obtained. The relative yields of all crops given by the various treatments were:

No liming	Burnt lime	Ground limestone		
		Coarse	Medium	Fine
56	111	106	110	113

It is evident that there is little to choose between the treatments and that it is not necessary to insist on very fine grinding of a limestone. The coarse material used contained 88 per cent between 20 and 40 mesh, while 93 per cent of the intermediate material was finer than 40 mesh, and 95 per cent of the finely ground sample passed through a 100 mesh sieve.

In another series on similar lines, burnt lime was compared with four different grades of ground limestone. Of the coarsest, corresponding to commercial limestone dust, only 22 per cent passed through a standard 100 mesh sieve; the next fairly coarse grade contained 35 per cent below 100 mesh; the third grade was comparable with many commercial forms of ground limestone at 52 per cent; and the fourth grade all passed through the 100 mesh sieve. With 11 crops on 5 sites the relative yields given by these treatments were:

No liming	Burnt lime	Ground limestone			
		Fineness of grinding			
		22%	35%	52%	100%
62	105	108	107	109	111

Again, very fine grinding was shown to be unnecessary.

Magnesian Lime

In a series of experiments on 26 crops burnt lime and ground limestone were compared with magnesian lime, both burnt and ground, all being given at equivalent rates of dressing. In this series, sugar beet and mangels benefited more from the magnesian materials, while barley seemed to do better with the ordinary high calcium lime. The relative yields on 100 sites were as follows:

No liming	High calcium burnt lime	High calcium ground limestone	Burnt magnesian lime	Ground magnesian limestone
40	112	114	118	116

The General Conclusions to be drawn from this work are:

1. The chief alternative forms of liming materials give very similar yields when used at equivalent rates.
2. Very fine grinding of limestone is unnecessary. Finely ground samples of ground limestone, of which 40 per cent pass the standard 100 mesh sieve, are as effective as the equivalent burnt lime.
3. Magnesian burnt lime and limestone are as effective as ordinary burnt lime and limestone at equivalent rates.

DURATION OF EFFECT

Except during the first few days after application, when burnt lime raises the pH of soil more quickly than do other forms, there is little to choose in speed of action between ground limestone or chalk and ground burnt lime. Ground limestone or chalk often has the advantage, because the amount to be spread, being twice that of burnt lime, makes more even application and incorporation possible. The maximum effect on soil pH occurs about four years later, and thereafter the pH decreases slowly. This is probably due to a gradual breakdown of the coarse particles of limestone or chalk, the finest (100 mesh) material being effective first.

The length of time during which the effects of an application of lime will last and the rate of loss depend upon rainfall, soil texture, organic matter, cropping, manurial treatment and soil pH (see page 6 for full details). The main points may be briefly summarized as follows.

Rainfall passes through the soil and gradually dissolves the lime compounds, which are thus lost in drainage water. This is the most important cause of soil acidity.

Because of the leaching effect of rain and soil drainage, lime always moves downwards in undisturbed soil and the surface becomes acid. An extreme instance of this may sometimes be seen on downland, where the surface soil may be a few inches only above solid chalk or limestone, and yet it becomes so acid that only acid-tolerant plants will grow.

The relative proportions of sand and clay, as well as of organic matter, in a soil greatly influence the retention of lime. Losses are most rapid from sands low in organic matter. Maintenance dressings should be given on these soils every three to four years, whereas on heavy soils the interval may be six to seven years. The length of interval must depend, however, on the lime status of the soil and the quantity of lime applied.

Heavy cropping, especially of fodder crops that take up much calcium from the soil and are not fed to stock on the field, steadily removes lime from the soil.

Some fertilizers react with the soil calcium to form soluble calcium salts, and similarly with magnesium, which may then be washed out of the soil. The best example is sulphate of ammonia, which reacts with lime to form calcium sulphate. Each cwt of sulphate of ammonia uses the equivalent of about 1 cwt ground limestone or chalk. This rate of loss can be borne without difficulty on land which either has large natural reserves of lime or receives regular maintenance dressings.

Soils of high pH with reserves of lime lose it more rapidly than very acid soils. Well-limed soils require a rather large 'bare maintenance' dressing than under-limed soils.

LIME SPREADING

Since liming materials differ in fineness of grinding, and soil particles are affected only by lime particles that are very close to them, it is most important that the time and method of application should be chosen to give the best distribution. The coarser grades need to be used at rather higher rates than the finer grades, and wet or lumpy material should obviously be used at still higher rates to allow for the difficulty of spreading evenly. Patchy crops on acid land that has been limed are often due to unevenness of distribution of lime, or liming too close to the date of sowing for proper incorporation to have taken place. Patchiness, due to differences in soil pH in one part of a field to another, also occurs because of local variations in soil conditions, such as drainage, depth and the underlying rock. These poor acid patches can be seen easily enough when crops are in the ground. If they are small, there is no point in trying to correct them specially, except on intensively cropped market gardens. Where, however, a quarter of an acre or more is distinctly more acid than the rest of the field, it is often worth while to give a heavier dressing of lime to the part that needs it.

On Arable Land

The depth of ploughing will affect the technique of applying lime. Most recommendations for liming are based on a depth of 6 in. of soil, but where deep ploughing is done care must be taken to allow for the extra soil to be treated, especially if the subsoil is much more acid than the soil near the surface.

Liming material should normally be applied to the surface of arable land, and not buried deeply. It should be cultivated in such a way as to keep it within reach of the roots of the young seedlings. If, however, the soil is very acid throughout or is deep ploughed, a full lime dressing may be ploughed down and another full dressing given to the newly exposed surface. Normal cultivation should then result in even corporation. When dunging is also to be done, there should generally be an interval of a few months between it and heavy liming.

A good point for the rotational application of lime is on the root break (except for potatoes). Occasionally, however, it may be necessary to lime for potatoes if the land is deficient in calcium, particularly if it has been recently reclaimed from moor or woodland.

On Grassland

Carbonate of lime (chalk or limestone) may be applied at any convenient time in the year. There are many acres of permanent grassland and of rough grazings which would benefit from liming, and the treatment of these during the summer, when the land will best stand the weight of heavy vehicles, relieves the rush of orders in the autumn and spring, and enables arable land to be dealt with at times when it is not occupied by crops.

Application of liming materials to grass should preferably be made during dry weather so that the particles may fall as completely as possible through to the soil. When the grass is wet the blades become coated with the lime which dries and sticks to them. Thorough working with a spiked or pitchpole harrow is helpful in order to work it quickly into the soil, especially if there is a mat of old herbage. It is advisable to remove stock from the fields for a day or two after liming, or until rain has washed the dusty particles off the grass. On very acid grassland scheduled for ploughing up, a first application should be well mixed into the old turf to help to rot it down and a second may be given after ploughing.

Machinery for Spreading

There are various methods of applying lime to the land, all designed to achieve as far as possible a uniform spread of the liming material over the field. In earlier days, when the main liming material was burnt lime, distribution over the field was done entirely by hand. Two main methods were employed: one was to slake the heap of lime at the entrance to the field and, when chemical action had ceased, to cart the lime over the field and spread by shovel from the back of the cart. The second method was to distribute the burnt lime in small heaps at regular intervals over the field and, when it had naturally slaked, to throw it around by shovel. The efficiency of both these types of spreading depended primarily on the labourer who undertook the task. The operation was unpopular, because of the dusty nature of the work and the risk of injury to the horse or men from incompletely slaked lime. These two methods are still the only effective ones for dealing with lump lime.

During the past decade or two, the shortage of farm labour and the understandable reluctance to handle caustic lime has increased the use of ground carbonates and screened chalks, and has led to the development of machines for spreading lime mechanically at standard rates per acre.

In the early days, a spinning plate was driven from the up-ended cardan shaft of an old lorry axle. This machine was trailed behind an ordinary delivery lorry and the lime, instead of being shovelled on to the land, was shovelled on to the spinning plate. The speed of the lorry and the regularity of the shovelling governed the rate and uniformity of spreading. This improvised device worked quite well and had the advantage of cheapness.

Later, specially built machines incorporated the spinning plate with the mechanical delivery of lime from a bulk hopper. Other methods of delivery have been devised, some based on the spinning plate, and others on the centrifugal action of one or more vertical spinners. Several types now in use are the direct outcome of these prototypes and can be roughly divided into self-propelled and tractor-drawn spreaders. For working in particularly hilly areas half-track and full-track vehicles have been adapted for use as lime spreaders.

The development of spreading by machine has led the producers of burnt lime, particularly in the north and west of England, to offer a roughly crushed burnt lime (kibbled lime) with a maximum particle size of about one inch, which is cheaper than ground burnt lime and suitable for mechanical spreading.

Another method of spreading has recently been tried, in which a finely powdered lime is fed into the airstream of a fast centrifugal fan and blown out over the area to be dressed. The results of further trials with this machine are awaited with interest.

The possibility of using aircraft for spreading lime on the more inaccessible hilly areas has been tested, and it seems possible that aircraft could be used effectively for delivering and spreading it in these areas. However, the problem of cost still remains to be examined thoroughly before any firm recommendations can be made. Consideration has also been given to the use of helicopters for transporting lime to inaccessible hilly areas, with a view to subsequent hand spreading.

DANGER OF OVER-LIMING

Crops can be seriously affected when too much lime of any kind is applied to soil. Excess lime alters the availability to crops or livestock of some of the minor nutrients (trace elements) and the balance of essential minerals taken up into herbage for livestock. Over-liming of some sands, marshland and other soils rich in organic matter encourages manganese deficiency in oats (Grey Speck) and in sugar beet (Speckled Yellows), while on other types of soils swedes develop Brown Heart and sugar beet Heart Rot, due to boron deficiency.

On soils in Somerset and Gloucestershire derived from the Lower Lias Clays, plants take up more molybdenum than is normal and this condition is aggravated by liming. Such pastures become in local phrase 'teart', so that cattle grazing there suffer from excessive and sometimes fatal scouring. When there is a risk of such troubles, advice should be sought before applying lime.

TECHNICAL ADVICE

The farmer is strongly urged to seek advice from the National Agricultural Advisory Service both on the amount of lime, if any, needed by the soil of each of his fields and also on the relative rates of dressing of different materials. In this way he gets to know which fields are in most urgent need and can arrange his liming programme accordingly. He is also saved the risk and expense of applying lime unnecessarily.

Lime in Horticulture

UNDER the general heading of horticultural crops is included a great variety of vegetables, fruits and flowers with widely differing requirements for lime. Many liming problems arise and the pH value of his soils is of great importance to the grower. The requirements of the various crops are briefly discussed below, but it may be said that on the average horticultural holding where a wide range of vegetables and other crops are grown, the aim should generally be to maintain a pH value of or about 6·5 or nearly neutral.

Ground limestone or chalk is a satisfactory and cheap liming material and has the advantage of being easy to store and pleasant to handle. There is no reason to believe that burnt or hydrated limes have any special virtues other than being slightly quicker in action, although hydrated lime is a long-established favourite among horticulturists. Burnt lime should not be used immediately before sowing or planting. Except in orchards, all forms of lime are best applied after ploughing and then thoroughly worked into the soil, although under strongly acid conditions there is something to be said for ploughing in part of the dressing of lime and cultivating the remainder into the surface.

Magnesian limestone has certain special uses in horticulture, which are discussed under the various crops below. It should be noted that ground magnesian limestone is produced in areas which are remote from the main horticultural districts and in these cases, transport costs may be heavy.

Too much lime in the soil can be just as harmful as too little and it is well known that micro-nutrient deficiencies may be induced on certain soils which have been over-limed or which naturally contain it in excess. Deficiencies of manganese and boron can be induced in this way and are briefly discussed under the various crops. An excess of lime is also to be avoided in that loss of organic matter from the soil is likely to be particularly high under alkaline conditions. In addition, calcium is antagonistic to the uptake of potash by the plant, and potash deficiency may be particularly severe on soils of high lime content.

VEGETABLE CROPS

For general market garden purposes a soil with an adequate reserve of lime is necessary, and a soil pH value of 6·5 may be regarded as desirable. Crops differ, of course, in their ability to grow satisfactorily at various levels of acidity, but it must be borne in mind that such a division is an arbitrary one, much depending on the soil and the rainfall. Crops often grow reasonably well in the wetter areas at levels of acidity at which they would fail in the drier parts of the country.

Levels of pH required for satisfactory cropping

About pH 6·5 (*need nearly neutral conditions*)

Beans (runner and French), beetroot, brassica crops (e.g. cabbage, cauliflower), carrots, celery, onions, leeks, lettuce, parsnips, rhubarb and spinach. (Note: Rhubarb tolerates quite acid conditions but probably grows best when well supplied with lime.)

About pH 6·0 (*satisfactory under slightly acid conditions*)

Broad beans, peas, radishes, sweet corn, swedes and turnips.

About pH 5·5 (*satisfactory under moderately acid conditions*)

Chicory, cucumbers, marrows, potatoes and tomatoes.

The following list gives pH values below which crops may be expected to fail, or at any rate be unsatisfactory in the eastern half of England. As already explained, however, the level at which failure may occur is greatly influenced by soil, rainfall and previous manuring (especially with farmyard manure which tends to offset the effects of acidity). Crops may sometimes be found growing reasonably well at lower pH values than those given.

Levels of pH below which crops may fail

pH Value	*Crops likely to fail*
6·5	Cauliflower, spinach.
6·0	Lettuce, leeks, onions, parsnips.
5·5	Beans (runner, French and broad), peas, beetroot, brassicas, carrots, celery.
5·0	Cucumbers, marrows, potatoes, tomatoes, sweet corn, rhubarb.

Of the crops listed, cauliflower and broccoli are especially liable to fail if the soil is at all acid. On some soils, for example, cauliflower and broccoli develop a strap-like condition of the leaves known as 'Whiptail', in which the plants are often blind and fail to produce a saleable curd. This condition is usually due to a deficiency of the micro-nutrient molybdenum which has been induced by soil acidity. Such cases can frequently be prevented by an adequate dressing of lime applied well in advance of the crop. Not all cases of 'Whiptail' are due to soil acidity and careful enquiry may be required in any specific instance to find the exact cause. Spinach is another crop which is particularly sensitive to acidity and under acid conditions plants remain stunted and yellow.

Many market gardening areas, for obvious reasons, are situated on light soils from which the loss of lime by leaching is considerable. This loss is often aggravated by the heavy dressings of fertilizers which are given and by the removal of crops from the land. It follows, therefore, that regular attention to liming is necessary on many market garden soils. Although it is quite impossible to give a reliable general guide, once the lime status of the soil has been raised to a satisfactory level, 1–2 tons per acre of chalk or ground limestone every few years should be a sufficient maintenance dressing. In order to raise the pH value to any desired level heavier dressings of lime are required on heavy soils than on light ones.

A few diseases of vegetables are affected by the lime content of the soil, of which the best known are Club Root of brassicas, and Common Scab of potato. Club Root is less serious on soils well supplied with lime, but Common Scab is usually worse on such soils.

Manganese and boron deficiencies can be induced by over-liming and are likely to be a problem to the vegetable grower on some soils. Manganese deficiency gives rise to a characteristic chlorosis (yellowing) of the older leaves of many vegetable crops, and in some varieties of peas brown spots are developed inside the seed, the so-called 'Marsh Spot'. Boron deficiency can also be a serious problem, more especially in dry seasons, giving rise to such conditions as Canker in beetroot, Brown Heart in swedes and Heart Rot in celery.

Magnesium deficiency is sometimes a problem in vegetable crops, particularly in cauliflowers, which exhibit a characteristic yellow marbling of the older leaves. On acid soils this deficiency can be controlled by dressings of ground magnesian limestone which supplies magnesium as well as correcting soil acidity.

GLASSHOUSE CROPS

The cost of liming for field crops can be an important part of the total production costs, and for economic reasons it may well be desirable to maintain soil pH at below optimum level in some circumstances. Under glass, the cost of liming is negligible in relation to other production costs, and border soils or potting composts are therefore always maintained at the *optimum* pH for plant growth, so far as possible.

Whereas field soils are subject to leaching by rainfall, soils under glass are watered artificially, and this has important effects on the lime content of the soil. On many holdings hard water is used, and the quantities of chalk applied to the soil during water application can result in a soil pH of over 7·0 which is maintained without the application of lime from other sources. Conversely, where the water is soft, leaching may be more severe than in field soils because of water applications in excess of those resulting from natural rainfall outdoors: this situation applies especially where plants are grown in pots or boxes. The lime content of the water supply to the holding has to be taken into account, therefore, in determining liming treatments.

For most of the common glasshouse crops the optimum soil pH range is 6·0 to 6·5. It is undesirable to allow the pH to move outside the optimum range unless there are practical circumstances such as those related to the water supply, which make precise control very difficult. For certain crops such as winter lettuce pH 6·0 should be considered the minimum safe level.

For more detailed information on the response of crops to soil pH levels reference should be made to the Ministry bulletins on specific glasshouse crops, but the following examples cover the problems associated with soil pH which are most frequently encountered in commerce:

Acid Soils

Extremely acid loams are sometimes used without lime addition for potting composts or bedding plant production, and at pH levels below 5·0 total failures often occur arising probably from a combination of lime deficiency and toxicities due to micronutrients such as aluminium and manganese. The low pH can be rectified by the addition of chalk, ground limestone, or magnesian limestone either to the loam when it is stacked before sterilization, or to compost during preparation. Table 5, page 42, indicates the quantities needed.

On certain types of soils manganese toxicity may occur after steam sterilization. This effect is most severe in acid soils, and can be rectified by increasing the lime and phosphate status of the beds. Manganese toxicity effects have been observed on tomatoes, lettuce, carnations and chrysanthemums, amongst other crops.

Alkaline Soils

At pH levels over 7·0, boron deficiency has been observed in tomato and carnation crops. Manganese deficiency may also occur in these conditions. Iron deficiency due to pH levels over 7·0 occurs commonly in glasshouse roses and can be severe. Primulas grown as pot plants are also subject to iron deficiency. Other pot plants such as azaleas and ericas are extremely intolerant of lime and require a soil pH of 5·0 or below and the same pH level is required for 'blueing' hydrangeas.

TABLE 5

Amount of lime added to loam at stacking

pH of loam	Approx. weight of powdered chalk or ground limestone (per 9 in. layer of stacked loam)	
	Loams, silty loams and silt loams	Silty clay loams and clay loams
	oz per sq. yd	oz per sq. yd
6·3	0	0
6·2	2	2½
6·1	4	5
6·0	6	7
5·9	7	9
5·8	9	12
5·7	11	14
5·6	13	16
5·5	15	18
5·4	16	21
5·3	18	23
5·2	20	25
5·1	22	28
5·0	24	30

FLOWER CROPS IN THE OPEN

There is considerable variation among flower crops in their tolerance of acidity. Many of the commoner flowers grown for cutting grow most satisfactorily at pH values between 6·0 and 6·5. Lupins grow well under quite strongly acid conditions and fail on soils with an excess of lime. Rhododendrons, azaleas, ericas (except *E. carnea* and certain other species) and many species of lilium and primula grown outdoors are also intolerant of lime and require soils with low pH values, although sometimes it seems as if soils can be too acid even for these plants. Hydrangeas are in a special class; in strongly acid soils (around pH 5·0) the flowers tend to be blue, whereas in soils with pH values between 6·0 and 7·0 they are pink or red. The key factor for 'blueing' is the availability of aluminium in the soil, which is influenced by its pH value, but the variety, and probably the phosphate and nitrate content of the soil, are also important. Hydrangeas are subject to iron deficiency, and do not grow well on soils of high lime content. Alkaline soils are unsuitable for general nursery work.

MUSHROOMS

In the production of mushrooms, some form of lime is commonly used in the soil covering or casing of the beds. Soils for casing need to be of suitable texture and in good physical condition, and a pH value between 7 and 8 is often recommended. With acid soils in need of lime, carbonate of lime (ground limestone or chalk) should be thoroughly mixed with the soil a short time before use. A suitable method is to build a heap of soil in 6 in. layers and to add to each layer carbonate of lime at the rate of about 1 lb per sq. yd of surface. After lime has been added to the last layer of the heap, the whole should be thoroughly turned to effect proper mixing. If the soil is to be heat sterilized, the addition of carbonate of lime should be deferred until after sterilization. It is inadvisable to use very acid soils for casing, even with the addition of the considerably larger dressings of lime which would be needed to reduce acidity.

Recently, various peat and chalk mixtures have been widely employed for casing. One such mixture consists of equal parts of medium to coarse peat and of chalk (½ or ¾ in. to dust).

FRUIT AND HOPS

The majority of fruit crops grow best in a slightly acid soil (about pH 6·0) and there does not appear to be any foundation for the old belief that the stone fruits, such as plums and cherries, require large amounts of lime. Excess of lime should be avoided since iron and manganese deficiencies can be serious problems in both top and soft fruits. On the other hand, strongly acid soils are also undesirable and fruit crops can be unsatisfactory under these conditions.

In grass orchards strong acidity is also undesirable for the sake of the sward. When lime is applied to cultivated orchards and soft fruit plantations it should only be lightly worked into the surface soil, taking care to avoid damage to the root systems of the trees or bushes.

Magnesium deficiency is commonly found in fruit crops on certain soils, especially in dessert apples and gooseberries; where the soil is acid, dressings of ground magnesian limestone will help to prevent or cure this deficiency.

Black currants differ from most of the fruit crops in requiring a soil with a good reserve of lime, and a pH value of 6·5 should be aimed at.

With the increased use of simazine for the control of weeds the acidity of the surface layer of soil may be increased and this encourages manganese toxicity. This trouble has been reported in young trees of 'Lord Lambourne' and 'Cox's Orange Pippin'.

Hops are fairly tolerant of soil acidity and liming should not be required unless the pH value falls into the range of 5·5 to 6·0.

Lime for Livestock

LIME plays an important part in livestock farming in two ways; indirectly as a factor in the production of grass and other fodder crops and directly as an essential nutritive additive in the diet, normally in the form of limestone flour or chalk.

SOIL LIMING

The indirect factor of application of lime to the soil is of significance only to ruminants, which may derive a considerable proportion of their feed from grazing during spring and from conserved grass or other forage crops as hay or silage during the winter, together perhaps with kale or other brassicas.

The diets of pigs and poultry under most modern systems of husbandry consist entirely of concentrates, and even for those which are kept under less intensive systems the overall contribution to their nutrition made by foraging is very small and may be disregarded in the context of this section.

The aim with grassland is to produce a maximum of grass of high feeding value, or crops of good quality hay or silage. This is achieved by a combination of good management and fertilizer application, linked with the important factor of correct lime status of the soil.

On land in good heart and of suitable pH the more desirable, prolific species of grass will tend to become dominant, giving rich leafy swards of high nutritive value. Acid conditions favour the growth of the poorer, less productive species, resulting in restricted growth rates of a herbage of generally inferior feeding value.

The benefits arising from the liming of grassland of this nature can be quite dramatic, with greatly increased yields of more nutritious and palatable feed, which in turn permits desirable increases in rates of stocking.

This increased productivity automatically places a burden upon the soil which in turn must be counteracted by suitable and adequate manuring if the benefits of the liming are to be maintained.

Liming can be applied too generously and in some circumstances may prove disadvantageous.

The excessive or unnecessary application of lime to grassland has been shown in some circumstances to be a factor in bovine infertility, apparently because of its effect, through the grass, upon the calcium/phosphorus metabolism in the cow. An unduly alkaline condition of the soil, brought about by excessive liming, may have the effect of suppressing the uptake of certain desirable 'trace' elements or of increasing that of molybdenum which is an undesirable one. Some of these are no more than possible effects since it is by no means easy to show conclusively that a particular disorder has arisen as the direct effect of a trace element deficiency. Molybdenum excess in herbage, resulting in the impairment of copper metabolism in the grazing animal with the usual effects of hypocuprosis, is an exception. Where there is a high molybdenum potential in the soil its uptake by the herbage will almost inevitably be aggravated by liming. Indiscriminate and excess liming is therefore a practice not to be encouraged.

Reports that the liming of grassland is followed by increased worm burdens in sheep, husk in young cattle and other parasites in livestock may be regarded as arising from management problems rather than directly attributable to the liming. Improved pastures invite more stock, and the rise in population will increase the risk of parasite infection.

LIME LOSSES

Removal of lime through crops sold off the land is discussed elsewhere in this bulletin, and the production and removal of stock and milk from grassland can also result in lime depletion. Such losses cannot be assessed with any accuracy because the picture is frequently complicated by supplementary feeding.

In many circumstances hand feeding of concentrates to animals on grass takes place, at least during part of the grazing season. Where such supplementary feed consists of mineralized concentrates, these will carry a 'safety excess' of calcium in order to correct for the utilization efficiency of the animal, which will be voided on to the field in the excreta. This would have to be set against that amount present in the ingested grass which was retained by the animal or secreted in the milk.

Part of the lime removed from a field in a hay or silage crop may well be returned later in the muck or slurry.

Estimates of lime removed from soil by way of stock or milk production can therefore only be made in the light of individual circumstances, and generalizations can be of little value.

Removal by milk produced solely from grass is approximately 40 lb limestone per thousand gallons of milk. A ton of hay will remove the equivalent of roughly 30 to 40 lb limestone and 5 tons silage some 50 lb. These amounts are comparatively small relative to the acreage of grass involved but they could amount to significant items over a succession of years, particularly on soils naturally low in calcium reserves.

MINERAL SUPPLEMENTS

The direct use of lime compounds in livestock production consists of inclusion in the concentrated foods, usually as limestone flour or chalk.

Cereals and vegetable protein foods which comprise the greater part of all compound animal feeds are all relatively high in phosphorus but low in calcium. They are therefore incapable of meeting fully the needs of the animals to which they are fed and are said to be 'out of balance' in respect of optimal ratios of calcium and phosphorus.

By making use of the knowledge of what these foods contain a better 'balance' is secured by the appropriate inclusion of limestone flour or chalk. Instances where correct ratios are achieved but overall insufficiency remains can be corrected by the inclusion of bone flour or other source of both calcium and phosphorus. In doing this, it is the concentrate diet which is adjusted in isolation, taking no account of what is contained in the supplementary bulk foods.

It would be impractical to attempt to adjust the diet of every animal in every set of circumstances, and for all practical purposes it suffices to maintain a reasonable balance of calcium and phosphorus in the concentrate feeds, with an overall safety surplus referred to earlier.

Pigs present no such problem since they are on all-concentrate diet which offers complete control. Limestone or chalk can be used to ensure the desired level of calcium.

In many respects poultry are the same, but the laying bird has a very large calcium requirement for egg shell production. This may be met in two ways, depending upon the system of housing and management of the laying flock. It is usual for birds in batteries to receive the whole of their calcium allowance in their mash or pellets, which may amount to 6 per cent of limestone flour. Some favour the same system for birds on deep litter or in other less restricted environments, but an alternative method is to include part of the limestone in the diet and to provide the remainder as limestone grit in hoppers from which the birds can help themselves.

This section gives only a brief and general outline of the uses of lime for livestock. The more detailed and fundamental aspects of the role of calcium in animal nutrition will be found in other publications.

Appendix

Regulations Under Fertilisers and Feeding Stuffs Act, 1926

New consolidated Regulations* under this Act came into force on 1 July, 1968 (except in relation to small packs of fertilizers on which the enforcement of the new requirements is deferred until 1 October, 1970). These Regulations do not introduce any new requirements which affect the sale of agricultural lime except that if certain additives are included these have now to be declared. The following particulars have been compiled from the Act and its regulations to assist readers who buy and sell agricultural lime.

Under Section 1(1) of the Fertilisers and Feeding Stuffs Act, 1926, every person who sells for use as a fertilizer of the soil any article included in the Second Schedule to the Act, whatever may be the name under which the article is sold, is required to give the purchaser on or before delivery or as soon as reasonably practicable thereafter a statement in writing (referred to in the Act as a 'statutory statement') containing the following particulars:

(*a*) the name under which the article is sold;

(*b*) such particulars (if any) of the nature, substance, or quality of the article as are mentioned in the second column of the Schedule.

This requirement, however, does not apply (1) to sales of two or more articles that are mixed at the request of the purchaser before delivery to him; and (2) to sales of small quantities (i.e., 56 lb or less) if the article sold is taken in the presence of the purchaser from a parcel or receptacle bearing a conspicuous label on which the above-mentioned particulars are marked.

It is provided in the Act that a statutory statement, or any voluntary statement of additional particulars of quality, given by a seller of any such article shall have effect as a written warranty by the seller that the stated particulars are correct. Further, in the case of articles not included in the Schedules, any voluntary statement has effect as a warranty.

Every purchaser of lime in respect of which a warranty has been given has the right under the Act to have a sample taken and analysed, and to be furnished with a certificate of the result of the analysis. This certificate will show him whether or not the article complies with the warranty. All that the farmer or other purchaser has to do is to (*a*) request the Official Sampler appointed by the County Council or County Borough Council, as the case may be, to take a sample, and (*b*) pay the prescribed fee, if any. The request should be made and posted promptly after the arrival of the consignment as the sample must be taken within fourteen days after delivery of the goods or receipt by the purchaser of the statutory statement or warranty, whichever date may be the later.

If the analysis discloses a deficiency or breach of warranty, the purchaser will decide whether to make a claim against the seller or not. No other action can follow, as no prosecution can be instituted by a Local Authority against a seller in respect of any deficiency where the sample has been taken on a farm by an Official Sampler. The name and address of the nearest Official Sampler can be obtained on application to the Clerk to the Local Authority.

* The Fertilisers and Feeding Stuffs Regulations, 1968 (S.I. 1968 No. 218), price 8s. 6d. (by post 9s.) and Amendment Regulations (S.I. 1968 No. 883), price 6d. (by post 9d.).

Particulars of Liming Materials

The table below has been compiled to include the implied definitions (column 2) of any article (column 1) included in the Fourth Schedule, the particulars required in the Statutory Statement under the Second Schedule (column 3), and the limits of variation permitted (column 4).

Explanatory Notes

Implied Definition (column 2). The definitions are as set out in the Fourth Schedule of the new regulations. Where an article is sold as a fertilizer and is described by a name defined in the Fourth Schedule, there is a warranty that the article conforms to the implied definition. In the case of every article the description of which includes the expression 'commercially pure', it is implied that no other matter may be added.

Particulars to be contained in the Statutory Statement. Except where otherwise stated, a Statutory Statement has to be given for any article described (column 1), under whatever name it may be sold or offered for sale and notwithstanding that it contains a substance not mentioned in this table.

The particulars (column 3) have to be given as a definite percentage of the weight of the article and not as a range of percentages (see Limits of Variation below).

Neutralizing Value (column 3) must be expressed in terms of calcium oxide (CaO).

The prescribed sieve (columns 2, 3 and 4) shall be the British Standard Test Sieve, mesh number 100.

Limits of Variation (column 4). The percentages under the limits of variation are percentages of the whole bulk.

Table of Liming Materials

Article (1)	Implied Definition in Fourth Schedule to the Regulations (2)	Particulars to be contained in Statutory Statement (3)	Limits of Variation (4)
Burnt magnesian lime, ground or otherwise	Commercial calcium and magnesium oxides containing more than 5·5 per cent of magnesium (Mg)	Neutralizing Value	Neutralizing Value, one-tenth of the amount stated
Burnt or quicklime, ground or otherwise	Commercial calcium oxide containing not more than 5·5 per cent of magnesium (Mg)	Neutralizing Value	Neutralizing Value, one-tenth of the amount stated
Calcium hydroxide; hydrated lime; slaked lime	The product obtained by slaking burnt lime	Neutralizing Value	Neutralizing Value, one-tenth of the amount stated
Chalk	Cretaceous limestone	None	

Article (1)	Implied Definition in Fourth Schedule to the Regulations (2)	Particulars to be contained in Statutory Statement (3)	Limits of Variation (4)
Chalk, ground	Cretaceous limestone which has been reduced in size so that it will pass through a sieve of $\frac{1}{4}$ inch square apertures	Neutralizing Value	Neutralizing Value, one-tenth of the amount stated
Chalk, screened	Cretaceous limestone that will pass through a sieve having apertures not exceeding 3 inches square	Neutralizing Value. Amount that will pass through a declared British Standard Test Sieve	Neutralizing Value, one-eighth of the amount stated. Amount that will pass through a declared British Standard Test Sieve, one-tenth of the amount stated
Limestone, ground	Sedimentary rock consisting largely of calcium carbonate but containing not more than 3 per cent of magnesium (Mg) which has been reduced in size so that 100 per cent will pass through a sieve of $\frac{3}{16}$ inch square apertures, not less than 95 per cent will pass through a sieve of $\frac{1}{8}$ inch square apertures and not less than 40 per cent will pass through a British Standard Test Sieve, Mesh No. 100	Neutralizing Value. Amount that will pass through a British Standard Test Sieve, Mesh No. 100	Neutralizing Value, one-twentieth of the amount stated. Amount that will pass through a British Standard Test Sieve, Mesh No. 100, one-twentieth of the amount stated
Magnesian limestone, ground	Sedimentary rock consisting largely of the carbonates of calcium and magnesium but containing more than 3 per cent of magnesium (Mg), which has been reduced in size so that 100 per cent will pass through a sieve of $\frac{3}{16}$ inch square apertures, not less than 95 per cent will pass through a sieve of $\frac{1}{8}$ inch square apertures and not less than 40 per cent will pass through a British Standard Test Sieve, Mesh No. 100	Neutralizing Value. Amount that will pass through a British Standard Test Sieve, Mesh No. 100	Neutralizing Value, one-twentieth of the amount stated. Amount that will pass through a British Standard Test Sieve, Mesh No. 100, one-twentieth of the amount stated
Mixed lime	A product, not being a by-product, or a mixture of by-products from manufacturing or other processes, obtained by mixing two or more of the forms of liming materials defined in this table	Neutralizing Value	Neutralizing Value, one-tenth of the amount stated
Slaked magnesian lime	The product obtained by slaking burnt magnesian lime	Neutralizing Value	Neutralizing Value, one-tenth of the amount stated

Index

(109339) Dd. 142440 K.40 3/69

MINISTRY OF AGRICULTURE
FISHERIES AND FOOD

Plant Pathology

A quarterly publication presenting original contributions on plant diseases, plant pests, rodent and bird damage, nutritional and physiological disorders of interest to the mycologist, entomologist, helminthologist, soil or nutrition chemist, meteorologist and plant physiologist

Single copies **7s. 6d.** (by post 8*s*.)

Yearly subscription **36s.**
(including postage)

Published by

HER MAJESTY'S STATIONERY OFFICE

and obtainable from the Government Bookshops in London (post orders to P.O. Box 569 S.E.1), Edinburgh, Cardiff, Belfast, Manchester, Birmingham and Bristol, or through any bookseller